ENVIRONMENTAL POLICIES FOR AGRICULTURAL POLLUTION CONTROL

ENVIRONMENTAL POLICIES FOR AGRICULTURAL POLLUTION CONTROL

Edited by

J.S. SHORTLE AND D.G. ABLER

Department of Agricultural Economics and Rural Sociology
Pennsylvania State University
USA

CABI *Publishing*

363.73945
E61

CABI *Publishing* is a division of CAB *International*

CABI Publishing
CAB International
Wallingford
Oxon OX10 8DE
UK

CABI Publishing
10 E 40th Street
Suite 3203
New York, NY 10016
USA

Tel: +44 (0)1491 832111
Fax: +44 (0)1491 833508
Email: cabi@cabi.org
Web site: http://www.cabi.org

Tel: +1 212 481 7018
Fax: +1 212 686 7993
Email: cabi-nao@cabi.org

A catalogue record for this book is available from the British Library,
London, UK.

Library of Congress Cataloging-in-Publication Data
Environmental policies for agricultural pollution control/edited by J.S. Shortle
and D.G. Abler.
  p. cm.
  Includes bibliographical references (p.).
  ISBN 0-85199-399-0 (alk. paper)
   1. Agricultural pollution--Government policy. 2. Water--Pollution--Government
  policy. 3. Environmental policy--Economic aspects. I. Shortle, J.S. (James S.)
  II. Abler, David Gerrard, 1960–

 TD428.A37 E58 2001
 363.739'45--dc21

2001020355

ISBN 0 85199 399 0

Typeset in Photina by Columns Design Ltd, Reading.
Printed and bound in the UK by Biddles Ltd, Guildford and King's Lynn.

Contents

Contributors   vii

Preface   ix

1. **Agriculture and Water Quality: the Issues**   1
   J.S. Shortle, D.G. Abler and M. Ribaudo

2. **Environmental Instruments for Agriculture**   19
   R.D. Horan and J.S. Shortle

3. **Voluntary and Indirect Approaches for Reducing
   Externalities and Satisfying Multiple Objectives**   67
   R.D. Horan, M. Ribaudo and D.G. Abler

4. **Estimating Benefits and Costs of Pollution Control Policies**   85
   M. Ribaudo and J.S. Shortle

5. **Non-point Source Pollution Control Policy in the USA**   123
   M. Ribaudo

6. **Policy on Agricultural Pollution in the European Union**   151
   N. Hanley

7. **Decomposing the Effects of Trade on the Environment**   163
   D.G. Abler and J.S. Shortle

References   183

Index   213

Contributors

D.G. Abler, Department of Agricultural Economics and Rural Sociology, College of Agricultural Sciences, Pennsylvania State University, 207 Armsby Building, University Park, PA 16802, USA

N. Hanley, Department of Economics, University of Glasgow, Adam Smith Building, Glasgow G12 8RT, UK

R.D. Horan, Department of Agricultural Economics, 87 Agriculture Hall, Michigan State University, East Lansing, MI 48824-1039, USA

M. Ribaudo, United States Department of Agriculture, Economic Research Service, 1800 M St NW, Room 4004, Washington, DC 20036-5831, USA

J.S. Shortle, Department of Agricultural Economics and Rural Sociology, College of Agricultural Sciences, Pennsylvania State University, 112 Armsby Building, University Park, PA 16802, USA

Preface

Environmental experts have recognized for years that agriculture is a major cause of water quality problems. Governments have taken up the problem with varying intensity and success. On the whole, progress has been limited and serious problems remain. There is now considerable support within environmental agencies and environmental groups for expanding pollution control efforts in agriculture. However, there is little consensus about appropriate policies.

As with other types of pollution, significant reductions in agriculture's contribution to water pollution will require the application of either enforceable regulatory approaches or changes in the economic environment such that farmers find it in their economic interest to adopt 'environmentally friendly' production practices. The appropriate choices among the range of options that fall within these boundaries are the subject of much debate. One option would be to pursue command and control regulations that require farmers to adopt certain management practices or technologies. This has been the dominant approach to air and water pollution control, but there are other options that are likely to have greater political and economic appeal.

There is a huge engineering and hydrological literature on managing water pollution from agriculture. This literature is an important resource for policy analysts, but many of the key questions that must be addressed in policy design and evaluation involve economic questions. What will it cost to reduce water pollution from agriculture, and how does that compare with costs of reducing pollution from other sources instead? What are the economic benefits from water pollution control? Why do the traditional approaches to water pollution control in agriculture, which involve education and voluntary compliance, have a limited impact? How will producers and agricultural markets respond to innovative policy initiatives? How will changes in agricultural markets induced by trade liberalization affect the location and severity of agriculture's contribution to pollution, and the costs and effectiveness of environmental policy choices? What are the merits of public spending on 'green' agricultural technology when competition for public resources is intense?

Our fundamental goal in this volume is to provide a resource for economists, other professionals and students interested in economic dimensions of designing and evaluating pollution control policies for agriculture. This volume includes chapters describing theoretical and empirical research on policy design, methods for policy evaluation, the policy experiences of various countries and linkages between agricultural trade and the environment. There are some sections that make use of mathematics and more advanced economic analysis, but on the whole we believe the book will be accessible to interested students, professionals and analysts with varied backgrounds.

Our thanks to our editor Tim Hardwick at CABI *Publishing* for his great patience in seeing us through to the end. We also thank Rose Ann Alters and Lubing Wang for various contributions to this project. We would also like to acknowledge the Economic Research Service of the US Department of Agriculture (Cooperative Agreement No. 43-3AEL-8-80058) for its support of our research on pollution control policies for agriculture and the collaborations it has enabled between the contributors to this book.

James Shortle
David Abler
December 2000

Chapter 1

Agriculture and Water Quality: the Issues

JAMES S. SHORTLE[1], DAVID G. ABLER[1] AND MARK RIBAUDO[2]

[1]Department of Agricultural Economics and Rural Sociology,
Pennsylvania State University, University Park,
PA 16802, USA; [2]USDA, Economic Research Service,
Washington, DC 20036-5831, USA

Agriculture has pervasive impacts on water quality. Conversion of forests, wetlands and prairies to crop and grazing lands has reshaped landscapes and the hydrology and ecology of agriculturally developed regions. Reservoirs constructed to supply water to agriculture and other sectors have destroyed significant natural assets by inundation.[1] Dams and diversions impede fish migration, alter stream flow regimes and water temperatures, and trap sediments. Consequences include severe degradation of aquatic and riparian habitats, with significant threats to aquatic species. Surface runoff from cropland carries salts, fertilizers, pesticides, pathogens and other pollutants into surface waters, damaging aquatic ecosystems and wildlife, degrading drinking water supplies, and impairing water for commercial and recreational uses. Pesticides and other chemicals applied to cropland also enter aquifers used for drinking water, posing risks to human and animal health.

Accessible and relatively comprehensive environmental assessment data make it easy to use the USA to illustrate the significance of these problems. At present, scores of species of fish are threatened or endangered in rivers where habitat has been degraded by large-scale irrigation projects (Moore *et al.*, 1996). Bird deaths, deformities and reproductive failures at the Kesterson Reservoir in California in 1983 highlighted the risks of agricultural drain water to wildlife. A recent assessment by the US Environmental Protection Agency (USEPA) concluded that agriculture is the leading cause of impaired water quality in rivers and lakes, and among the leading causes of impaired estuaries and shorelines in the US (USEPA, 2000a). Pesticides and nitrates are routinely found in groundwater supplies in agricultural regions of the country (USGS, 1997, 1999a).

These problems are by no means unique to the US. Similar problems are found in developed and developing countries, and they are drawing increasing attention from scientists, environmental advocates and policy-makers (FAO, 1996; OECD, 1998). Concerns for the environmental impacts of dam building have become a significant barrier to water development for agriculture and other uses in many countries. Increasingly, scientists and environmental groups are calling for the transference of irrigation water reserved for agriculture to the support of freshwater ecosystem services (Postel, 1999). Protecting ground- and surface water quality from chemicals leached from farm fields and polluted agricultural runoff has become a major water quality policy issue in many countries. Agriculture in the United States and in the European Union (EU) has become a major target of regulatory initiatives to protect water quality (OECD, 1998). Our interest in this volume is the choice of policies to reduce agriculture's contribution to ground- and surface water pollution. While mounting evidence indicates that agriculture is a major cause of water quality problems in many regions, policy measures to control pollution loads from agricultural activities have generally been quite limited in comparison with measures that have been taken to control other sources of water pollution. Large 'point sources' of pollution and manufacturers of environmentally harmful products such as pesticides have been the main targets of regulatory programmes.[2] Agriculture and other 'non-point sources' have largely escaped direct regulation. This imbalance is now recognized as limiting progress towards water quality goals in regions where agriculture, or other non-point sources, are a significant cause of water quality problems. Furthermore, the failure to extend pollution controls to agriculture and other non-point sources increases the costs of water quality protection by precluding efficient allocation of control between point and non-point sources (Freeman, 1990).

There is considerable support within environmental agencies and environmental groups for expanding pollution control efforts in agriculture. However, there is little consensus about appropriate policies. There has been a heavy reliance on voluntary compliance approaches in the past but these are generally acknowledged to have had limited impact. As with other types of pollution, significant reductions in agriculture's contribution to water pollution will require the application of either enforceable regulatory approaches or changes in the economic environment such that farmers find it in their economic interest to adopt 'environmentally friendly' production practices. The appropriate choices among the range of options that fall within these boundaries are the subject of much debate.

The principal objectives of this volume are to develop an understanding of the basic types of instrument that can be used to control non-point source agricultural pollution, to lay out the economic and political issues involved in choosing between instruments, and then to assess the fundamental advantages and disadvantages of these alternative instruments. We also consider the linkages between agricultural trade and the environment, and between agricultural development and the environment. These two broader issues are currently at the forefront of the debate about agriculture and the environment and are also part of the debate over international trade and globalization.

In short, given that policy-makers want to 'do something' about agriculture and the environment, we attempt to provide insight based on economic research about what should be done. As much experience shows, laws and bureaucracies can acquire a life of their own as interest groups of one sort or another spring up with a stake in maintaining the status quo, making it very difficult to modify or repeal unsound legislation. Thus it is important in the area of non-point source agricultural pollution, as in any area, to 'do it right' the first time. Poor policy choices to limit agricultural pollution can be quite expensive and show little in the way of environmental benefits. On the other hand, research suggests that good choices can accomplish much at relatively low cost.

Agricultural Pollutants and Their Impacts

Agricultural pollutants include sediment, nutrients (nitrogen and phosphorus), pesticides, salts and pathogens. While farmers generally do not intend for these materials to move from the field or enterprise to water resources, they often do. For example, as much as 15% of the nitrogen fertilizer and up to 3% of pesticides applied to cropland in the Mississippi River Basin make their way to the Gulf of Mexico (Goolsby and Battaglin, 1993). In the paragraphs that follow we present an overview of agricultural pollutants and the kinds of damage that they can cause. While examples of these agricultural impacts can be found in many countries, there is a woeful lack of systematic information in most countries on water quality generally and on agriculture's impacts in particular. Marc Ribaudo presents a detailed look at agriculture and water quality in the USA in Chapter 5 and Nick Hanley does the same for Western Europe in Chapter 6.

Nutrient pollution

One of the leading water quality issues associated with agriculture in developed countries is nutrient pollution by nitrogen and phosphorus.[3] Nutrients, chiefly nitrogen, potassium and phosphorus, are applied to cropland in manufactured fertilizers and animal manures to increase yields. In areas with intensive live-stock production, such as The Netherlands and in many areas of the United States, manure may be applied to cropland primarily to dispose of the waste and only secondarily as a fertilizer.

Nutrients can enter water resources in four ways. *Runoff* transports pollutants over the soil surface by rainwater, melting snow, or irrigation water that does not soak into the soil. Nutrients move from fields to surface water while dissolved in runoff water or adsorbed to eroded soil particles. *Run-in* transports chemicals directly to groundwater through sinkholes, porous or fractured bedrock, or poorly constructed wells. *Leaching* is the movement of pollutants through the soil by percolating rain, melting snow, or irrigation water. Finally, nitrogen can enter water resources through *atmospheric deposition* (e.g. in rain).

Nitrogen and phosphorus are the nutrients of particular concern for water quality. Nitrogen in the form of nitrate is easily soluble and is transported in runoff, in tile drainage, and with leachate. Phosphate is only moderately soluble and, relative to nitrate, is not very mobile in soils. However, erosion can transport considerable amounts of sediment-adsorbed phosphate to surface waters. If soils have been over-fertilized, rates of dissolved phosphorus losses in runoff will increase due to the build-up of phosphates in the soil.

In a process known as eutrophication, increasing nitrogen and phosphorus levels in slow-moving waters stimulate algae growth and the resulting effects on the aquatic ecology can be dramatic. As algae bloom, they take up dissolved oxygen, depleting the oxygen available for fish and other aquatic life. They can also block the sunlight needed by aquatic vegetation, causing the vegetation to die off. This loss in vegetation then moves up the food chain, leading to the death of fish and other aquatic life. Eutrophication of fresh water is usually due to phosphates, while nitrates are usually the cause of coastal water eutrophication. Both nutrients tend to be important in the eutrophication of estuaries.

A number of recent assessments of water quality problems by national and international organizations point to the fact that eutrophication is not a trivial issue. Human activities have more than doubled the amount of nitrogen in the environment globally from 1960 to 1990, with the use of synthetic fertilizers accounting for more than half of that growth (National Research Council, 2000). In the US, the National Oceanographic and Atmospheric Administration (NOAA) conducted the National Estuarine Eutrophication Survey from 1992 to 1997 to assess the quality of the country's 138 major estuaries. The survey found that 44 estuaries (40%) exhibited high expressions of eutrophic conditions caused by nutrient enrichment (Bricker *et al.*, 1999). These conditions occurred in estuaries along all coasts, but were most prevalent in estuaries along the Gulf of Mexico and Middle Atlantic coasts. Human influences (point and non-point source nutrient pollution) were associated with 36 of the 44 estuaries. A recent USEPA report identifies nutrients as the leading cause of the impairment of lakes, and third most important cause of impairment of rivers and streams (USEPA, 2000a). Agriculture is listed as the primary source of pollutants causing impairments in both cases. The Organization for Economic Cooperation and Development (OECD) reports that in both rural and urban areas of Western European countries, the majority of nitrogen pollution is due to agricultural activities, although agriculture is a much smaller contributor to phosphate pollution.[4] The OECD also notes that concentrations of nitrates at the mouths or downstream frontiers of rivers in Western Europe and North America are generally increasing.

Apart from reducing biodiversity, the loss of aquatic life from eutrophication can cause significant aesthetic and economic damage. The growth and subsequent decomposition of algae can be unsightly and generate foul odours, an obvious disamenity for those living or working near polluted waters. The productivity of commercial fisheries can be reduced, affecting the economic welfare of people in the fishing industry and related industries. To the extent that the prices of fish and seafood are increased, consumers are also harmed. Recreational fishing, boating and swimming can be adversely affected, to the detriment of those who engage in these activities and those who earn their living from them.

There are growing concerns in many countries over the impact that concentrated animal feeding operations have on water quality and other rural amenities. As a result of domestic and export market forces, technological changes and industry adaptations, animal production industries have seen substantial changes over the past decade. There has been an expansion in the number of large confined production units and geographical separation of animal production and feed production. Indeed, the bulk of US egg and chicken production occurs today under factory-style mass production conditions. One consequence of this is 'hot spots', or regions with extremely high concentrations of animals and surplus animal wastes. Examples include the Chesapeake Bay and Albemarle and Tar-Pamlico Sound regions of the eastern US, most of The Netherlands, and eastern parts of Quebec. The geographical concentration of feeding operations can overwhelm the ability of a watershed to assimilate the nutrients contained in the waste and maintain water quality. In addition, the size and number of animal waste storage lagoons increase the chance of a leak or a catastrophic break. These potential risks were underscored in 1999 by the impacts of Hurricane Floyd in North Carolina (USGSb, 1999).

A leading case of nutrient pollution is found in the world's largest estuary, the Chesapeake Bay. The Chesapeake Bay is one of the most valuable natural resources in the United States. It is a major source of seafood, particularly highly valued blue crab and striped bass. It is also a major recreational area, with boating, camping, crabbing, fishing, hunting and swimming all very popular and economically important activities. The Chesapeake Bay and its surrounding watersheds provide a summer or winter home for many birds, including tundra swans, Canada geese, bald eagles, ospreys and a wide variety of ducks. In total, the Bay region is home to more than 3000 species of plants and animals (Chesapeake Bay Program, 1999).

Elevated levels of nitrogen and phosphorus in Chesapeake Bay have led to a severe decline in highly valued fish and shellfish in recent decades. For 1985, 77% of the nitrogen and 66% of the phosphorus were estimated to be from non-point sources (Chesapeake Bay Program, 1995). Agricultural non-point sources were by far the most important, contributing 39% of the nitrogen and 49% of the phosphorus. Twenty-seven per cent of the nitrogen was atmospheric non-point pollution, with 11% falling directly on the water. Nitrogen oxides from fossil fuel combustion were the primary source of the atmospheric nitrogen. In the United States in total, more than half the nitrogen emitted into the atmosphere from fossil fuel-burning plants, vehicles and other sources is deposited in US watersheds (Puckett, 1995). The shares of total nitrogen load to selected eastern US estuaries from atmospheric deposition have been estimated to range between 4 and 80% (Valigura *et al.*, 1996).

Another important case of nutrient pollution is found in the Northern Gulf of Mexico, where an oxygen-deficient 'dead' zone has more than doubled to 8000 square miles (20,720 km^2) since 1993 (National Science and Technology Council, 2000). The primary cause is believed to be increased levels of nitrates carried to the Gulf by the Mississippi and Atchafalaya Rivers, and a major source of nitrates is fertilizers and animal wastes from the Upper Mississippi Basin (Goolsby *et al.*, 1999). Agriculture is estimated to be the source of 65% of the nitrogen entering the Gulf from the Mississippi (Goolsby *et al.*, 1999).

Nitrates in drinking water supplies obtained from either surface water or groundwater can pose human health risks.[5] One disease caused by ingestion of nitrates is methaemoglobinaemia, better known as blue-baby syndrome because bottle-fed infants less than 6 months old are particularly susceptible. The disease, which causes a reduction in the ability of blood to supply oxygen to the body, can lead to death. The incidence of this disease is unknown, but it is considered to be rare in North America and Western Europe. Nitrates are also suspected as a cause of cancer. They react with other chemicals in the body to form N-nitrosamines, which are known to cause cancer in laboratory animals. However, there is no known relationship between human cancer and these compounds. Exposure to nitrates in drinking water is chiefly a concern to those whose source water is groundwater, which generally has higher nitrate concentrations than surface water (Mueller *et al.*, 1995).

The health risks associated with nitrates prompted the World Health Organization (WHO) to issue drinking water standards for nitrates more than two decades ago. The WHO standard of 50 mg of nitrates (NO_3) per litre is widely accepted and has been incorporated into law in the EU. The US and Canada have a somewhat stricter standard of 10 mg of nitrogen (N) per litre.[6] Given the large safety margins included in drinking water standards and the uncertainty about health effects, the threat to public health may not be too serious. However, regardless of the true magnitude of the risk, communities are required to comply with established drinking water standards and this can be very costly.

Pesticides

Pesticide use in crop protection began in the late 19th century, but the water-mark of modern pesticide use was the introduction of synthetic organic pesticides in the mid-1940s. These chemicals offered farmers a cheaper and more effective way of protecting crops than traditional methods such as weeding with machinery or by hand. The use of pesticides in agricultural protection in North America and Western Europe has increased dramatically since then.[7]

Like nutrients, there are a variety of possible fates for pesticides applied to fields and orchards.[8] Pesticides dissolved in runoff water or attached to eroded soil particles may be washed into streams, rivers, lakes and estuaries. Pesticides may also evaporate into the air or leach into groundwater. A particular pesticide's fate depends on many factors, including its chemical and physical properties, the method of application, soil characteristics and the weather. Pesticides can also find their way into water resources via direct application to control aquatic weeds, wind drift, or overspray from aerial applications. The cleaning of application equipment or disposing of unused products into wells can also pollute water resources.

Pesticide residues reaching surface water systems may harm freshwater and marine organisms, damaging recreational and commercial fisheries (Pait *et al.*, 1992). Pesticides washed into lakes, rivers and estuaries can lead to fish kills, and numerous cases have been documented. Aquatic species and their predators can suffer chronic effects from low levels of exposure to pesticides over prolonged periods. Pesticides can also accumulate in the fatty tissue of animals such as shellfish to levels much higher than in the surrounding water, and consumption of these animals may lead to chronic effects in predators. Moreover, the effects can be 'biomagnified' as 'bioaccumulated' pesticides are passed up the food chain. This is what made DDT so damaging, leading to its ban. Herbicides and insecticides can kill the plants and insects upon which birds and other wildlife feed.

Pesticides in drinking water supplies may also pose risks to human health. Some commonly used pesticides are probable or possible human carcinogens (Engler, 1993). The overall state of knowledge about chronic effects on human health is quite limited, but concern has been raised about the consequences of low exposures over long periods of time. One cause for this concern is the fact that farmers and farmworkers involved in the handling, mixing and application of pesticides tend to have a higher incidence of lung cancer and other types of cancer.[9] In addition to cancer, questions have been raised about other possible effects of pesticide exposure. For example, two nematocides found in groundwater, EDB and DCBP, were cancelled by the USEPA because they might cause genetic mutations and reproductive disorders, as well as cancer.

Regulation requires additional treatment by public water systems when certain pesticides exceed health-safety levels in drinking water supplies. As with nitrates, water supply systems can incur significant treatment costs when water supplies are contaminated.

Sedimentation and turbidity

Disturbing the soil through tillage and cultivation and leaving it without vegetative cover increases the rate of soil erosion. Dislocated soil particles can be carried in runoff water and eventually reach surface water resources, including streams, rivers, lakes, reservoirs and wetlands. Sediment causes various types of damage to water resources and to water users. Accelerated reservoir siltation reduces the useful life of reservoirs. Sediment can clog roadside ditches and irrigation canals, block navigation channels and increase dredging costs. By raising stream beds and burying streamside wetlands, sediment increases the probability and severity of floods. Suspended sediment can increase the cost of water treatment for municipal and industrial water uses. Sediment can also destroy or degrade aquatic wildlife habitats, reducing diversity and damaging commercial and recreational fisheries. In the United States, sediment is the leading cause of impairment of rivers and streams, with agriculture being the major source (USEPA, 2000a).

Sediment also is a delivery mechanism for phosphorus and other pollutants. Many toxic materials can be tightly bound to clay and silt particles that are carried into waterbodies, including some nutrients, agricultural chemicals, industrial wastes, metals from mine spoils and radionuclides (Osterkamp *et al.*, 1998). When sediment is stored, the sorbed toxins are also stored and become available for assimilation.

Mineral damage

When irrigation water is applied to cropland, a portion of it runs off the field into ditches and flows back to a receiving body of water. These irrigation return flows may carry dissolved salts, as well as nutrients and pesticides, into surface waters or groundwater. Increased concentrations of naturally occurring toxic minerals, such as selenium and boron, can harm aquatic wildlife and degrade recreational opportunities. As noted above, the risks of agricultural drain water to wildlife were highlighted by bird deaths, deformities and reproductive failures at the Kesterson Reservoir in California in 1983. Increased levels of dissolved solids in public drinking water can increase water treatment costs, force the development of alternative water supplies, and reduce the life spans of water-using household appliances. Increased salinity levels in irrigation water can reduce crop yields or damage soils so that some crops can no longer be grown.

While discussing the impacts of irrigation return flows, it is important to mention that diversion of water for irrigation is also a significant cause of water-related environmental problems. As we noted earlier, in the arid western United States, scores of species of fish are threatened or endangered in rivers where habitat has been degraded by large-scale irrigation projects (Moore *et al.*, 1996). Water diversion to agriculture and polluted irrigation return flows have contributed to the environmental disaster in the Aral Sea in Central Asia (Tanton and Heaven, 1999). On a lesser scale, fish are endangered by irrigation-related habitat degradation in the Murray–Darling River system in Australia (Postel, 1999).

Concerns for the environmental impacts of dam building have become a significant barrier to water development for agriculture and other uses in many countries. Increasingly, scientists and environmental groups are calling for the transfer of irrigation water reserved for agriculture to the support of freshwater ecosystem services. In the United States, these calls have led to changes in federally operated irrigation projects. For example, in 1992, the Central Valley Project Improvement Act allocated 800,000 acre-feet of water to ecosystem maintenance (USDI, 2000). The Murray–Darling Basin Commission in Australia has also taken steps to limit water withdrawals to protect fish habitats. Several countries in the Aral Sea Basin have agreed in principle that the sea itself ought to be regarded as an independent claimant to water resources in order to support the sea's ecosystems (Postel, 1999).

Pathogen damage

The problems of pathogen-contaminated water supplies is attracting increased attention (Olson, 1995; NRAES, 1996). In the United States, bacteria are the second most common cause of impairment of rivers and the major cause of impairment of estuaries (USEPA, 2000a). Potential sources include inadequately treated human waste, wildlife, and animal operations. Animal waste contains pathogens that pose threats to human health (CAST, 1996). Microorganisms in livestock waste can cause several diseases through direct contact with contaminated water, consumption of contaminated drinking water, or consumption of contaminated shellfish. Bacterial, rickettsial, viral, fungal and parasitic diseases are potentially transmissible from livestock to humans (CAST, 1996). Fortunately, proper animal management practices and water treatment minimize the risk to human health posed by most of these pathogens. However, protozoan parasites, especially *Cryptosporidium* and *Giardia*, are important etiological agents of waterborne disease outbreaks (CDC, 1996). *Cryptosporidium* and *Giardia* may cause gastrointestinal illness, and *Cryptosporidium* may lead to death in immunocompromised persons. These parasites have been commonly found in beef herds, and *Cryptosporidium* is estimated to be prevalent in dairy operations (USDA APHIS, 1994; Juranek, 1995).

On-farm versus off-farm environmental impacts

We have focused above on impacts of agriculture on off-farm water resources, but many of the activities that produce these impacts also cause on-farm damages. For example, soil erosion can cause on-farm productivity damages as well as off-farm water quality damages.[10] Irrigation can bring with it the well-known problems of salinization and waterlogging. Salinization occurs primarily as a result of the deposition of harmful salts contained in irrigation water around the root zones of crops, preventing them from absorbing needed water and nutrients.[11] Waterlogging prevents roots from penetrating the soil, also cutting off needed nutrients. The use of pesticides and fertilizers creates health risks for farmers and their families (Stokes and Brace, 1988).

The distinction between on-farm and off-farm environmental impacts is important when considering whether an environmental impact is a social problem warranting collective action, or a private problem. Provided that agricultural land markets work well, the costs of on-farm environmental impacts that affect the productivity of farmland are internal in that they will be capitalized into the value of the farmland (e.g. McConnell, 1983; Miranowski and Hammes, 1984; Ervin and Mill, 1985; Gardner and Barrows, 1985; Palmquist and Danielson, 1989). Accordingly, property owners (and owner-operators) will have a personal interest in mitigating practices that diminish the value of their asset. Similarly, provided farmers have good information about health risks from pesticides and fertilizers and information on safe handling practices, they can choose practices consistent with the level of risk they are willing to accept (e.g. Beach and Carlson, 1993; Antle and Pingali, 1994; Hubbell and Carlson, 1998; Ready and Henken, 1999).

The off-farm environmental impacts of agricultural production are an entirely different matter (Shortle and Miranowski, 1987). In the absence of government policies of one sort or another, individual farmers' contributions to these impacts do not show up on their own bottom line. They are external to the farm operation. Farmers may be as concerned about the environment as anyone else, or even more concerned, but if changes in farming practices to protect the environment are costly, then it may be asking a great deal to expect farmers to reduce their own incomes voluntarily for the sake of protecting the natural environment. This is particularly true when they have no reason to believe that their fellow farmers will follow suit and when they are very uncertain (and possibly sceptical) about the effects of their individual choices on the quality of water resources (Tomasi *et al.*, 1994). If there is a consensus that off-farm impacts need to be addressed, then collective action of one sort or another is clearly required (Baumol and Oates, 1988).

Do efforts by farmers to limit the on-farm impacts of soil erosion or reduce health risks from pesticides benefit off-farm water resources? In some cases, they will. For example, measures to reduce soil loss to conserve soil productivity should limit damages from sediment. However, farmers' self-protection is not always good for the environment. For example, no-till farming practices help to conserve soil but may result in increased use of pesticides (Fawcett *et al.*, 1994; USDA ERS, 1997). The use of large volumes of irrigation water to flush salts from root zones, thus reducing on-farm productivity losses from salinity, can increase the salinity of runoff, and thus the off-site damages (National Research Council, 1993). Further examples as well as a general discussion of conflicts between self-protection and the external cost of pollution can be found in Shogren (1993).

The off-farm impacts discussed above are summarized in Table 1.1. For

Table 1.1. Major impacts of agriculture on the environment.

Environmental problem	Description/manifestations of problem	Proximate cause(s)	People harmed
Eutrophication	Algae blooms Death of fish and other aquatic life	Runoff of nitrates from fertilizers Runoff of nitrates from animal manures Phosphates from fertilizers – via soil erosion	Workers in commercial fishing industry Consumers of fish and seafood Workers in water recreation industries Users of water recreation
Sedimentation and turbidity	Death of fish and other aquatic life Fills up rivers, bays, harbours and other bodies of water Clogs water treatment plants	Soil erosion	Workers in commercial fishing industry Consumers of fish and seafood Workers in water recreation industries Users of water recreation Residential and industrial water users Victims of flooding
Drinking water contamination	Nitrates in surface water and groundwater Pesticides in surface water and groundwater	Runoff and leaching of nitrates from fertilizers Runoff and leaching of nitrates from animal manures Phosphates from fertilizers – via soil erosion	General public in contaminated areas

each environmental problem, Table 1.1 lists its proximate cause(s) and the people who are harmed. The latter is critical from a political point of view because policy-making experience has shown that reducing environmental damages is much more feasible politically if those damages have identifiable, significant impacts on people. There is much less public interest in the 'environment' as an abstract entity or in preserving the environment for the sake of non-human animal species or plant species. Table 1.1 does not by any means constitute an exhaustive list of the impacts of agriculture on the environment, but it does list what we believe to be the most important impacts in developed countries from a policy-making point of view.

Options for Reducing Agricultural Pollution

Farmers can take many steps to reduce loadings of agricultural pollutants to water resources (Hrubovcak *et al.*, 1999). However, the availability of technological solutions only helps to define what is possible, not what is optimal. What is optimal will depend on the answers to some fundamental economic questions:

1. How should responsibility for pollution load reductions be allocated between agricultural and other sources?

In watersheds in which agriculture is the only source of water pollutants, or perhaps the only source of pollutants that cause a particular type of water quality impairment, reducing pollution loads from agriculture is the only option for achieving water quality objectives. However, agriculture is typically one of many sources of pollutants. For instance, nutrients entering the Chesapeake Bay in the United States originate from urban, suburban and agricultural runoff, industrial and municipal point source discharges, rural septic systems and atmospheric deposition. In such cases, decisions must be made about which sources to control and to what degree. Because the costs of reducing pollution can vary greatly from one source to another, these choices can have a large impact on the costs of water quality protection. Cost-effectiveness considerations suggest allocating greater responsibility to sources with lower control costs, and in our analysis we will focus on this criterion. However, legal, political and fairness issues cannot be disregarded and often must take precedence in policy-making.

2. How should responsibility for agricultural load reductions in a watershed be allocated among alternative farms?

Within a watershed, physical factors (e.g. location relative to streams, soil types, slopes, etc.) that influence water quality impacts of agricultural practices vary from farm to farm. This means that different farms have different impacts on water quality and that the costs of pollution control vary across farms. Cost-effectiveness considerations will suggest allocating greater responsibility to farms with lower control costs. Again, however, legal, political and fairness issues cannot be disregarded.

3. What types of environmental policy instrument should be used to achieve the desired outcomes?

Achieving water quality goals requires the choice and implementation of policy instruments that will lead private decision-makers to adopt pollution prevention and/or abatement practices that are consistent with public objectives. In the end, it is the choice of instruments that will determine the environmental and economic outcomes. 'Command and control' instruments have been the dominant approach to environmental policy in developed countries (Opshoor *et al.*, 1994). These instruments generally involve mandated use of specific pollution control technologies, or adherence to input restrictions, product standards, emissions quotas or other regulations. However, there is growing interest in the use of economic incentives and market-based approaches that have the potential to achieve environmental quality goals at lower costs than command and control instruments (Opshoor *et al.*, 1994; Anderson *et al.*, 1997).

In agriculture, environmental policy instruments – command and control or otherwise – have been rare. Regulations on pesticide use, and bans on certain pesticides, are examples of the command and control approach in agriculture. By and large, however, the emphasis in agriculture has been on voluntary compliance approaches that combine public persuasion with technical assistance to encourage and facilitate adoption of environmentally friendly technologies (OECD, 1989, 1993a). Assessments generally indicate that these programmes have had limited impact. While there are many reasons for this, economic research suggests that costs are a significant barrier to the adoption of environmentally friendly practices (Feather and Amacher, 1993; Dubgaard, 1994; Norton *et al.*, 1994; Feather and Cooper, 1995). Interest in alternatives, both traditional command and control approaches as well as economic incentives and market-based approaches, is growing as the limitations of the voluntary compliance approach become increasingly evident.

4. How do agricultural price and income policies complement or conflict with agricultural water quality policies? Alternatively, how can farm income and environmental goals be best reconciled?

A number of economic studies indicate that price supports, input subsidies and other agricultural policies influence the nature, size and spatial distribution of agricultural externalities through effects on the scale and location of production, input usage and structure (e.g. Lichtenberg and Zilberman, 1986; Abler and Shortle, 1992; Antle and Just, 1993; Weinberg *et al.*, 1993a,b; Liapis, 1994; Swinton and Clark, 1994; Platinga, 1996; Abrahams and Shortle, 2000). For example, policies that increase producer prices without restricting output (e.g. price floors, output subsidies, import restrictions) encourage farmers to increase production. Adverse environmental impacts occur in so far as these policies induce farmers to produce on environmentally sensitive lands and in so far as farmers make more intensive use of environmentally harmful inputs (e.g. pesticides, fertilizer, irrigation water, fossil fuels).

Agricultural policies that increase livestock production also imply an increase in the volume of livestock wastes. Similarly, input subsidies can have adverse impacts when they encourage the use of potentially harmful agricultural inputs.

One might think that supply controls such as production quotas or acreage restrictions would be environmentally beneficial because they limit agricultural output, but this need not be the case. For example, acreage restrictions may lead to substitution of environmentally harmful inputs such as fertilizers and pesticides for land. Moreover, output quotas may be environmentally harmful when the rents they create encourage production in environmentally sensitive areas. These types of policy conflict indicate that negative agricultural externalities can be reduced by agricultural policy reforms, and they have stimulated considerable interest in coordinating agricultural and environmental policies (OECD, 1989, 1993a).

Traditionally, agricultural policies have attempted, with varying degrees of success, to achieve objectives related to farm income, agricultural prices and agricultural trade (Gardner, 1990). Agricultural externalities, although influenced by the scale, location and methods of agricultural production, were at most a secondary consideration. However, a shift in priorities is evident in many countries. Agricultural policy is increasingly concerned with encouraging the supply of positive agricultural externalities and decreasing the generation of negative externalities (Ervin and Graffy, 1996; Poe, 1997).

5. At what levels (national, regional, local) and through what agencies (environmental, agricultural) of government should actions to reduce agricultural loading take place?

The efficiency of a policy designed to reduce agricultural pollution is affected by not only the instruments selected, but also at what level of government the policy is implemented. Is it best for a central authority to design incentives or to impose standards for desirable management practices, or is it best left to local authorities?

A basic principle of the economic theory of federalism is that economic efficiency in the provision of a public good is generally best served by delegating responsibility for the provision of the good to the lowest level of government that encompasses all of the associated costs and benefits.[12] In the case at hand, the public good is environmental quality. The assumption underlying this principle is that policy choices consistent with the collective preferences of the affected group are more likely when made by decision-makers who represent their interests. National regulatory policies may not be sensitive to local conditions. For example, the Clean Air Act and Safe Drinking Water Act in the United States call for the USEPA to set uniform maximum allowable levels of pollutants for the entire country. Similarly, the US Federal Water Pollution Control Act sets uniform minimum surface water quality standards and technology-based maximum effluent standards for industrial and municipal point sources. In contrast, it is generally in the political interests of local authorities to develop policies that are sensitive to local preferences.

Current point source controls constrain the freedom of sub-national authorities to devise policies (in both goals and means) that correspond to local costs and benefits. Economic criticism of uniform command and control policies is centred on the costs imposed by uniformity and the limited use of information about local conditions in devising local solutions. The approach leaves little latitude to allocate pollution abatement among alternative sources within the point source category or between the point and non-point categories to minimize costs. Both the uniform command and control policy instruments and national water quality goals can lead to pollution control levels in which the incremental costs exceed the incremental benefits at particular sites.

While much can be said in favour of decentralization, there are several dimensions of water pollution control that call for national involvement. For instance, there are problems for which uniformity of outcomes and centralized decision-making are efficient. National pesticide regulation is an example. National registration/cancellation policies are a crude instrument for addressing the range of societal issues associated with the use of pesticides. However, nationwide cancellation is an optimal decision when the expected marginal damage to human health and the environment from the use of pesticide is so large that it always exceeds the marginal benefit. USEPA pesticide decisions to date appear to fit this characterization fairly well (Lichtenberg, 1992; Cropper *et al.*, 1992). Centralized responsibility for decisions of this type reduce decision-making costs and thereby improve the cost-effectiveness of environmental protection.

A number of additional arguments for national or even international involvement can be advanced:[13]

- In many cases the impacts of non-point source pollution are most pronounced close to their point of origin. Contaminated groundwater does not move far from pollution sources. Lakes and small reservoirs are generally affected by local land uses. Likewise, streams and small rivers are impacted by land uses within relatively small watersheds. The impacts of agricultural runoff on water quality are generally most pronounced in small lakes and reservoirs, and small rivers (Goolsby and Battaglin, 1993). However, in cases where pollutants spill over from one jurisdiction to another, optimal policies for upstream jurisdictions should take into account the benefits that are received in downstream jurisdictions. This accounting is unlikely with decentralized approaches for the simple reason that the political fortunes of upstream decision-makers depend on the preferences of upstream voters but not downstream voters. Spillovers of agricultural pollutants are known to occur and in some cases contribute significantly to downstream problems. Leading examples are the Chesapeake Bay and Gulf of Mexico problems mentioned above. Moreover, even when the physical impacts of pollution occur entirely within a given political boundary, costs may still spill over into other jurisdictions. There is ample evidence in the economic literature that people are concerned about resources that they do not use themselves (Fisher and Raucher, 1984). These non-use values are especially important in the case of unique natural resources. Indeed, some evidence suggests that the high level of concern for water quality that is found in public opinion surveys is often a concern for the quality of water where others live and work rather than one's own water (Bord *et al.*, 1993).

- Policies that reduce water pollution from agriculture can provide external benefits beyond those associated with downstream water quality improvements. Specifically, policies that reduce the production of nationally subsidized commodities and the use of nationally subsidized inputs would reduce deadweight losses associated with tax distortions of labour and capital markets (e.g. Lichtenberg and Zilberman, 1986; Alston and Hurd, 1990; Weinberg and Kling, 1996). However, while society gains, the benefits will be largely external to the sub-national units of government. As with the downstream benefits from water quality improvements, such decision-makers have little or no incentive to consider these benefits in sub-national policy-making.

- Economically efficient pollution policies require information about the demand for water quality, the linkages between water quality and economic activity, and the costs of changes in economic activity to reduce pollution. The information intensity of non-point pollution control, which we will discuss later, is generally an argument in favour of watershed-based approaches and therefore decentralized planning. Watershed-specific information may have some value for research and other purposes but the primary value will be for local planning and administration. The costs of obtaining the information should therefore be allocated largely to the specific watershed. If the net benefits of watershed-specific plans developed using watershed-specific information are no more than the net benefits of centralized plans developed without the benefits of watershed-specific information, then the case for decentralized planning, including cooperative plans for problems involving spillovers, is weakened. This would be the case only in the unlikely event that the value of watershed-specific information is zero.

- In addition to supporting the case for decentralized management plans, information needs also provide support for national research and development. Some types of information needed in non-point pollution control planning cut across watersheds and localities. Examples include the impacts of nitrate or pesticide ingestion on human health, hydrological principles governing the fate and transport of pollutants, and the basic economics of evaluating alternative techniques and the merits of alternative policy approaches. Information of this type is optimally provided at the national level.

An issue that is receiving increasing attention in the implementation of agricultural pollution programmes is which government agencies should be responsible. Environmental agencies have traditionally been responsible for protecting water quality, but it is not uncommon for agricultural agencies to be responsible to some degree for agricultural pollution control programmes. The belief is that these agencies are most familiar with farmers' production practices and individual situations. Many state agriculture departments in the US already have programmes that provide financial incentives and technical assistance for conservation practices, and agricultural environmental programmes are often modified versions of these programmes. The obvious problem is one of regulatory capture. Agricultural agencies, whose mission has been traditionally to promote the interests of agricultural producers and food processors, may elevate the well-being of farmers, particularly those represented by powerful lobbies, above the timely achievement of environmental goals (Browne, 1995). A recent study by the Environmental Law Institute (ELI, 1997) finds that, in the US,

> Integration of technical assistance and cost share with enforcement has been difficult in some respects. Many agriculturally-oriented agencies do not want to be associated with enforcement. The case studies show that even states with the most fully developed enforceable mechanisms generally seek to assure that in addressing agriculture and forestry, the enforcement function is assigned to a separate entity from the cost-share and technical assistance function.

Overview

The remaining chapters take up, to varying degrees, the issues outlined above. In Chapter 2, Horan and Shortle outline major environmental instruments for agriculture and review the specialized theory of non-point pollution control that has emerged to help to guide choices. They also examine empirical results on the performance of alternative approaches. Horan, Ribaudo and Abler examine 'indirect' approaches to reducing pollution from agriculture in Chapter 3. These include education programmes, research and development, and steps to reduce or eliminate conflicts between farm income support policies and policies to reduce water quality impacts. Ribaudo and Shortle discuss empirical methods for analysing environmental policies for agriculture in Chapter 4. The North American and European experiences in controlling water pollution from agriculture are examined by Ribaudo and Hanley, respectively, in Chapters 5 and 6. Abler and Shortle examine the coordination of environmental policies for agriculture with agricultural trade policies in Chapter 7.

Endnotes

1. Approximately 70% of water withdrawals worldwide are to supply agriculture (FAO, 2000).

2. Pollution sources are generally distinguished as point or non-point according to the pathways the pollutants or their precursors follow from the place of origin to the receiving environmental media. Pollutants from point sources enter at discrete, identifiable locations. Industrial facilities that discharge residuals directly into air or water from the end of a smoke stack or pipe exemplify this class. Pollutants from non-point sources follow indirect and diffuse pathways to environmental receptors. Open areas such as farm fields, parking lots and construction sites from which pollutants move overland in runoff into surface waters or leach through the permeable layer into groundwaters are examples. The classification of pollution sources as point or non-point is not always clear cut or fixed. For further discussion, see Shortle and Abler (1997).

3. For more information on surface water pollution generally see OECD (1986), Cooper (1993), Smith *et al.* (1993) and USEPA (2000a,b).

4. See the OECD series *The State of the Environment* and the companion series *Environmental Data Compendium*. See also the OECD's *Environmental Performance Reviews* for several countries, including Germany (OECD, 1993b) and The Netherlands (OECD, 1995).

5. For more information on the health risks from nitrates, see Cantor and Zahn (1988) and Mirvish (1991). For more on groundwater pollution from nitrates and pesticides in the US, see Nielsen and Lee (1987), Spalding and Exner (1993) and USDA ERS (1994). For the OECD generally, see the citations in endnote 4.

6. The US–Canadian standards convert to 45 mg l^{-1} in terms of NO_3.

7. For statistics for the US, see USDA ERS (1994) and Osteen and Szmedra (1989). For the EU, see Brouwer *et al.* (1994). For OECD countries generally, see the citation in endnote 4.

8. For more information on pesticides and the environment, see OECD (1986) and Pimentel *et al.* (1991).

9. See Cantor *et al.* (1988). For a global overview of this issue, see the World Health Organization (1990).

10. See Wischmeier and Smith (1978). For statistics for the US on the distribution of highly erodible cropland and on-farm productivity losses from erosion, see USDA ERS (1994). The on-farm productivity losses from erosion in many developing countries are substantially higher than in developed countries (see Pimentel, 1993).

11. Salinization can also occur as a by-product of waterlogging in so far as waterlogging leads to a rise of saline groundwater to the root zones of crops. Salinization is common in arid and semi-arid regions of the world where there is not enough rainfall to leach salts from the soil.

12. Underlying the principle that responsibility for externalities ought to be allocated to the lowest level of government that encompasses the associated costs and benefits is the assumption that public decision-makers act in the public interest and weight costs and benefits to those involved more or less equally. This view corresponds to the traditional 'rational planning' or 'problem-solving' approach to public policy in economics. In this tradition, government serves the public interest and intervenes to correct problems associated with imperfect or missing markets. However, modern theories of public policy and regulation recognize that what governments do in fact is the outcome of the interplay between influential interest groups.

13. These arguments follow and expand on Shortle (1996).

Chapter 2

Environmental Instruments for Agriculture

RICHARD D. HORAN[1] AND JAMES S. SHORTLE[2]

[1]*Department of Agricultural Economics, Michigan State University,*
East Lansing, MI 48824-1039, USA;
[2]*Department of Agricultural Economics and Rural Sociology,*
Pennsylvania State University, University Park, PA 16802, USA

This chapter[1] examines the question of how to induce farmers who cause water quality damages through their choice of production practices to adopt pollution prevention and pollution control practices that are consistent with societal environmental quality objectives. The question is partly behavioural in that it requires an understanding of how production and pollution prevention and control practices will change in response to policy initiatives. Relevant actors include implementation agencies and participants in the markets in which farmers purchase inputs and sell products, as well as farmers themselves (Segerson, 1996; Davies and Mazurek, 1998; Russell and Powell, 2000). The question also requires understanding of the biophysical relationships between farming practices and water quality. However, evaluation of the merits of alternative approaches cannot be limited to their ability to induce improved environmental performance. The performance of the instruments with respect to other societal interests must also be considered when evaluating the options. These interests include various economic and non-economic criteria (Bohm and Russell, 1985; Segerson, 1996; Davies and Mazurek, 1998; Russell and Powell, 2000). Economic criteria include the social costs of control, including public sector administration costs, incentives for environment-saving technical change and flexibility in the face of exogenous change. These aspects are emphasized. Non-economic criteria include non-intrusiveness in private decision-making, political acceptability and fairness in the burden of costs and benefits.

The chapter begins with a look at some basic questions that must be addressed when choosing a strategy for agricultural non-point pollution controls, and that help to explain features of the agricultural pollution control problem that make economically and politically appealing policy choices difficult. The remainder of the chapter is devoted to the results of theoretical and empirical research on the design of pollution control instruments for agriculture and other non-point sources.

Some Fundamental Questions

Selecting strategies to reduce the environmental impacts of agricultural production requires choices about who must comply (e.g. all farmers, some subset, agricultural chemical manufacturers, etc.), how their compliance, or performance, will be measured, and how to induce changes in behaviour.

Whom to target?

It seems obvious enough that policies to reduce pollution from agriculture and other non-point and point sources ought to be directed at those who are responsible – the decision-makers who choose what to produce and how to produce it. However, assigning responsibility for non-point pollution loads to individuals is typically not easy. Routine metering of pollutant flows from individual farms is prohibitively expensive and often technically infeasible. For example, nitrogen fertilizer applied to cropland can have a variety of fates depending on how and when it is applied, weather events during the growing season and other factors. These fates include consumption by plants, leaching through the soil into groundwater, removal in surface runoff, or volatilization into the atmosphere. None of these fates, especially those that involve losses to different environmental media, are easily measured or predicted. Nor can individuals' contributions be routinely inferred from ambient concentrations in environmental media, because the latter are determined by the joint contributions of many unmeasured sources (both natural and anthropogenic).[2]

Uncertainty about who is responsible and the degree of responsibility creates significant problems for non-point pollution policy design.[3] Regulation of producers who cause little or no problem creates costs without offsetting benefits. Yet failure to target broadly enough diminishes effectiveness and limits opportunities for cost-effective allocations. Fairness is clearly an issue when producers are required to undertake costly activities in the public interest but there is uncertainty about whether the public interest is served at all. Similarly, political support for policies to regulate sources that may cause no actual or apparent environmental damage may be difficult to muster. The fairness and political considerations associated with uncertainty about the responsibility of individuals may help to explain the frequent use of 'moral suasion' and 'government pays' approaches to agricultural pollution control rather than 'polluter pays' approaches (Chapters 3, 5 and 6).

It is important to note that answers to the 'whom to target?' question, as applied to agricultural non-point sources, need not be limited to farmers. Other plausible answers are manufacturers of chemical inputs and providers of services such as fertilizer and pesticide application. This option is best illustrated by laws regulating the pesticides that chemical companies may offer on the market to farmers and other users. Intensive regulation of a comparatively small number of chemical manufacturers is easier, politically and administratively, than intensive direct regulation of the many households and small businesses that actually cause environmental harm through their activities and use of pesticides.

What to target?

What is it that society wants polluters to do? Here again the answer seems obvious enough. It is to reduce or limit the amount of pollutants that they release into the environment. This suggests that the measure of polluter performance, and the basis for regulatory compliance, ought to be polluting emissions. Not surprisingly, then, economic research on pollution control instruments has established that the economically preferred base for the application of regulatory standards or economic incentives is the flow of emissions from each source into the environment – provided that discharges can be metered with a reasonable degree of accuracy at low cost (e.g. Oates, 1995). As discussed above, this condition is not a characteristic of diffuse, non-point pollution problems in which pollutants move over land in runoff or seep into groundwater rather than being discharged from the end of a pipe. With pollution flows being for practical purposes unobservable, other constructs must be used to monitor performance and as a basis for the application of policy instruments.[4] The economics of designing policy instruments for agricultural and other non-point pollution externalities is therefore complicated by the fact that choices must be made between alternative bases as well as between types of regulation or incentive (Griffin and Bromley, 1982; Shortle and Dunn, 1986). Economically and ecologically desirable candidates will be more or less: (i) correlated with environmental conditions, (ii) enforceable and (iii) targetable in time and space (Braden and Segerson, 1993).

With instruments based on metered discharges eliminated from the non-point choice set, perhaps the obvious next choice is emissions proxies (e.g. estimates of field losses of fertilizer residuals to surface water or groundwater) or other farm-specific environmental performance indicators that are constructed from farm-specific data.[5] In the simplest cases, emissions proxies could be the use of polluting inputs, such fertilizer or pesticide applications by a farmer. More sophisticated indicators aggregate over inputs and other variables. One of the best known examples is the Universal Soil Loss Equation (USLE) developed by Wischmeier and Smith (1987) for predicting gross soil loss from cropland. Another widely used performance indicator is the difference between nutrient inflows and outflows in farm products (National Research Council, 1993; Breembroek *et al.*, 1996). For example, charges for Dutch livestock producers are to be assessed based on surplus phosphates from manure (Breembroek *et al.*, 1996; Weersink *et al.*, 1998).

Other options for bases are inputs or farming practices that are correlated with pollution flows (e.g. use of polluting inputs such as fertilizers, pesticides, irrigation water; the use of practices such as conservation tillage, integrated pest management). Still another option that has received considerable interest from economists is the *ambient concentrations* of pollutants in water resources. Inputs and emissions proxies are alternative measures of environmental pressures of agricultural production, while ambient environmental conditions are an environmental state variable.

What stimulus?

In addition to choices about whom to target and how to measure their performance, a catalyst is needed to get producers to undertake changes to improve their environmental performance. The least intrusive method is public persuasion combined with technical assistance to facilitate changes in behaviour. This approach has found extensive use in agriculture. For example, the USDA NRCS Conservation Technical Assistance (CTA) Program has provided technical assistance since 1936. By itself, public persuasion and technical assistance have had limited effects (Chapters 3, 5 and 6). However, technical assistance may be somewhat effective as a component of other programmes, such as the USDA NRCS Environmental Quality Incentives Program (EQIP).

A more direct stimulus is direct regulations (i.e. product, design or environmental performance standards) applied to farmers' choices of inputs and production and pollution control practices. Pesticide registration, which restricts pesticides available to farmers and sets conditions of use, is an example. Alternatively, farmers' decisions could be shaped through the use of economic incentives. Major options are taxes or liability for damages to discourage environmentally harmful activities, subsidies to encourage pro-environment behaviours, tradeable permits to ration environmentally harmful activities, and contracts in which environmental authorities purchase specified pro-environmental actions.

The tool kit

Table 2.1 combines the mechanisms with alternative compliance measures discussed above to define a range of instruments for agricultural non-point pollution control. The table is not meant to provide a comprehensive listing of all possible mechanisms that are used in practice or that have been proposed in theory. Instead, it is focused on those types of policy mechanism that have

Table 2.1. Pollution control instruments for agriculture.

Mechanism	Inputs/practices	Compliance measure	
		Emissions proxies	Ambient concentrations
Taxes/subsidies	Charges on fertilizer or pesticide purchases	Charges on modelled nutrient loadings	Ambient taxes
	Charges on manure applications	Charges on nutrient applications in excess of crop needs	
	Cost-sharing or other subsidies for inputs or practices that reduce pollution	Charges on estimated net soil loss	
	Cropland retirement subsidies		
Standards	Pesticides registration	Restrictions on modelled nutrient loadings	
	Restrictions on fertilizer application rates	Regulations on nutrient applications in excess of crop needs	
	Mandatory use of pollution control practices		
Markets	Input trading	Estimated emissions trading	
Contracts/bonds	Land retirement contracts		
	Contracts involving the adoption of conservation or nutrient management practices		
Liability	Negligence rules		Strict liability/negligence rules

received significant attention in the economic literature and which are described in this chapter.

There are many examples of product and design standards. Pesticide registration is the principal method for protecting the environment, workers and consumers from pesticide hazards in developed countries, and is used increasingly in developing countries (OECD, 1986; Dinham, 1993). Standards governing the amount and timing of manure applications and restrictions on the numbers of farm animals are used to control ammonia, phosphorus and nitrogen pollution from agriculture in The Netherlands (Broussard and Grossman, 1990; Dietz and Hoogervorst, 1991; Leuck, 1993).

Economic incentives applied to inputs and practices are also a dimension of agricultural non-point water pollution control programmes. Australia, Canada, Denmark, Sweden and the United States provide subsidies for adoption of pollution control practices in agriculture and some other sectors. Many of these subsidies come in the form of contracts, under which producers contract with a government agency to implement a negotiated set of practices for a specified time interval in return for payments. Major examples in the US currently include the USDA NRCS EQIP and Conservation Farm Option (CFO) programmes (USDA ERS, 1997). Subsidies are also offered at the extensive margin of production (i.e. to change land use) – for example, the US payments offered for shifting land to activities with lower environmental hazards. The major programme in the US is the Conservation Reserve Program (CRP), which contracts with farmers and pays them to convert land from row crop production to grassed cover or other uses. The total water quality benefits of the CRP when fully implemented have been estimated at nearly $4 billion (1988 dollars) (Ribaudo, 1989). Florida has offered a dairy herd buy-out programme as part of efforts to reduce nutrient pollution of Lake Okeechobee.

Taxes are also used to varying degrees. Typically, input taxes (in agricultural and other contexts) are levied at such low rates that they offer little incentive to reduce input usage (OECD, 1994a). The purpose is more often to generate revenue for environmental programmes than to reduce input use (Weersink *et al.*, 1998). For example, Iowa levies taxes above and beyond the usual sales taxes on fertilizers to raise money for conservation programmes, but the rates are not sufficient to produce much in the way of environmental impacts (Batie *et al.*, 1989).

Subsidies and regulations are the dominant mechanisms for reducing agricultural pollution. Subsidies are often used to reduce the costs of complying with mandated activities. In such cases, they serve primarily to spread costs and increase the political acceptability of direct regulations. However, there is considerable economic evidence that input-based incentives could be effective in bringing changes in resource allocation. Firms respond to changes in the costs of inputs, increasing the use of those that become relatively cheap, while conserving on those that become relatively expensive (Shumway, 1995). In the long run, input price responsiveness is even greater, as technologies are developed and adopted to conserve further on more expensive inputs and expand the use of those that are cheaper (Hayami and Ruttan, 1985). Thus policy-makers, if they are willing to use taxes or subsidies at levels that will have an impact, can expect results from input-based incentives.

Pollution trading, in which individual sources of pollution are provided with limited rights to pollute and allowed to trade these rights in markets, is drawing significant interest as a means for agricultural non-point pollution control in the US, with a number of pilot programmes under way or on the drawing boards. In most of these programmes, municipal and industrial sources of pollution that are regulated by the US National Pollution Discharge Elimination System (NPDES) are able to avoid costly discharge reductions at their own facilities by paying agricultural sources to reduce their emissions. Liability rules of law have also found application in the US, particularly for managing hazards from pesticides and other harmful chemicals (Wetzstein and Centner, 1992; Segerson, 1995).

The remainder of this chapter examines issues in designing different instruments, and their economic merits. The focus is on instruments that are enforceable, and that have the sole purpose of reducing agricultural non-point pollution. (Chapter 3 examines voluntary approaches and approaches that explicitly serve multiple objectives.) Our presentation roughly parallels the evolution of the theoretical economic literature on non-point instruments. Growing recognition of the magnitude of agriculture's contribution to the water pollution problems in the USA and Europe in the 1970s stimulated economic interest in the design of environmental policy instruments for reducing polluting runoff and groundwater contamination from agricultural production. This literature has focused on the three questions (whom to target, what to target, and what stimulus) that are raised above. Initially, researchers looked to the theoretical and empirical literature of environmental economics for guidance. Because this literature highlighted the control of conventional point sources of pollution where emissions are often readily observed, the emphasis there was on the economic merits of alternative types of discharge-based economic incentive (e.g. discharge charges/standards, discharge reduction subsidies, transferable discharge permits). As we noted above, the literature on discharge-based environmental instruments is of limited relevance to the design of pollution control instruments for agriculture (and other non-point sources) because the movement of pollutants from farm fields in runoff or through soil into drains or aquifers generally cannot be so measured. With unobservable pollutant flows, other constructs must be used as performance standards and as a basis for the application of policy instruments.

Griffin and Bromley (1982) was the first in a series of three particularly influential articles, the others being Shortle and Dunn (1986) and Segerson (1988), that initiated the development of an economic theory of non-point pollution control. Griffin and Bromley and Shortle and Dunn focused on the design of instruments that require measurement of farmers' choices of inputs or practices, either because the instruments are input based (e.g. fertilizer taxes), or because farm inputs are used to construct an emissions proxy. The sections below begin with instruments of this type, followed by the ambient-based instruments proposed by Segerson (1988), and closely related liability rules, which shift performance monitoring from farms or other enterprises that cause polluting emissions to the resources that are damaged. We then take up point–non-point trading systems and other recent developments.

Incentives and Regulations for Inputs, Practices and Emission Proxies

While pollution flows from farm fields are not easily or cheaply measured, hydrological process and statistical models have been developed to assess these flows given measurements of appropriate land characteristics, weather and farm production practices (Chapter 4). Griffin and Bromley (1983) proposed that environmental decision-makers use information on the relationship between production choices and emissions provided by such physical models, which they refer to as a non-point production function, for direct measurement of emissions. They described how the information could be used to construct economically efficient input tax/subsidy schemes or input standards. They also described the construction of an economically efficient tax and standard for an emissions estimate obtained using the non-point production function to map from farm production inputs to emissions. Their model (with some modifications that we introduce) and results provide a good starting point from which to gain insights into the complexities of designing cost-effective non-point pollution control instruments.

Consider a watershed in which pollution from both point and non-point sources contributes to water quality impairments in a body of water such as a lake or estuary. For simplicity and because of our focus on agriculture, denote non-point sources as farms (although in principle these sources could include non-agricultural sources as well). Denote the non-point production function for the ith farm ($i = 1, ..., n$) by $r_i(x_i, \alpha_i)$, where r_i is non-point emissions, x_i is a $(1 \times m)$ vector of production and pollution control choices (inputs), and α_i represents site characteristics (e.g. soil type and topography).[6] This function is Griffin and Bromley's non-point production function, which represents a perfect estimate of the unobservable non-point emissions. Because it is an estimate (albeit a perfect one), the non-point production function is a non-point emissions proxy. In contrast, point source emissions, denoted e_k for the kth point source ($k = 1, ..., s$), are observable without error. For simplicity, we often refer to non-point emissions as runoff to distinguish them from point source emissions. Ambient pollution concentrations in receiving waters are expressed as a function of point and non-point emissions, and natural background levels of the pollutant, ζ, and watershed characteristics and parameters, ψ, i.e. $a = a(r_1, ..., r_n, e_1, ..., e_s, \zeta, \psi)$ ($\partial a/\partial r_i \geq 0 \ \forall i$, $\partial a/\partial e_k \geq 0 \ \forall k$). We refer to the relation $a(\cdot)$ as the fate and transport function.

Environmental damage costs, D, are an increasing function of the ambient pollution concentration, i.e. $D = D(a)$ ($D' > 0$). Damages occur because ambient pollution levels (and hence water quality) affect the ability of the water resource to provide economic services. For example, changes in a waterbody's characteristics (fish species present, fish abundance, physical appearance) can affect its value for recreation. Similarly, suspended sediment, algae and dissolved chemicals may increase the cost of providing water for industrial and municipal use.

Finally, assume for simplicity that polluters are risk-neutral, profit maximizers, and that they do not have any collective influence on the prices of inputs or outputs. In this case, the change in producers' quasi-rents (profits less fixed costs) is an appropriate measure of the costs of pollution controls (Just *et al.*, 1982; Freeman, 1993). Denote the *i*th farm's expected profit for any choice of inputs by $\pi_{Ni}(x_i)$. This function can be thought of as restricted profit function where the restrictions apply to inputs that enter the non-point production function (Shortle and Abler, 1997). Similarly, point source profits, restricted on emissions, are denoted by $\pi_{Pk}(e_k)$.

The model described above can be used to design optimal instruments based on estimated emissions and input use, but only after adopting a decision criterion to guide instrument design. In a purely economic approach to pollution control policy, we would have as an objective minimizing the social costs of pollution and its control or, equivalently, maximizing the difference between the expected benefits of polluting activities and the expected costs of the resulting pollution. In this case, the objective function would be

$$SNB = \sum_{i=1}^{n} \pi_{Ni}(x_i) + \sum_{k=1}^{s} \pi_{Pk}(e_k) - D\left[a\left(r_1, \ldots, r_n, e_1, \ldots, e_s, \zeta, \psi\right)\right]$$

Alternatively, separating the choice of environmental quality objective from the economic problem of how best to achieve it, a least-cost, or cost-effective, allocation minimizes the social costs of achieving the pre-specified objective (Baumol and Oates, 1988). The environmental objective could take a variety of forms. In terms of the variables contained in our model, possibilities would include limits on emissions (in total, by sector, etc.), limits on the ambient concentration, or a limit on the economic damages costs. The real-world options are much broader. Given that environmental consequences (both physical and economic) depend on the ambient concentration, we choose a constraint limiting an unspecified indicator of environmental degradation, $W(a)$, to be no more than a target T defining the maximum acceptable level of degradation. The indicator is continuous and increasing in the ambient concentration. The performance measure W is quite general and could take on a variety of forms, including $W = D$ and $W = a$. The least cost control problem is then

$$\max NB = \sum_{i=1}^{n} \pi_{Ni}(x_i) + \sum_{k=1}^{s} \pi_{Pk}(e_k)$$

$$s.t. \, W(a) \leq T$$

Given the limits on the information available for benefit–cost analysis, and the salience of non-economic criteria for choices of environmental quality objectives, cost-effectiveness is arguably the more useful economic concept for policy analysis and is the criterion we adopt for instrument design.

The first order conditions for an interior, cost-effective solution are[7]:

$$\frac{\partial \pi_{Ni}}{\partial x_{ij}} - \lambda \frac{\partial W}{\partial a} \frac{\partial a}{\partial r_i} \frac{\partial r_i}{\partial x_{ij}} = 0 \quad \forall i, j \tag{1}$$

$$\frac{\partial \pi_{Pk}}{\partial e_k} - \lambda \frac{\partial W}{\partial a} \frac{\partial a}{\partial e_k} = 0 \quad \forall k \tag{2}$$

where λ is the shadow value associated with the environmental constraint. Equation (1) requires at the margin that the gain in profits from the use of any input on any farm equals the environmental opportunity costs. The environmental opportunity costs are the input's marginal contribution to the environmental degradation indicator multiplied by the shadow cost of the environmental constraint. Similarly, equation (2) requires at the margin that the gain in profits from emissions equals the environmental cost of emissions. The solution to (1) and (2) is denoted by $x_{ij}^* \ \forall i,j$, $e_k^* \ \forall k$, and λ^*.

In the absence of environmental policy, farmers and point sources will maximize profits by equating the marginal profits associated with their choices to zero, i.e. $\partial \pi_{Ni}/\partial x_{ij} = 0 \ \forall i,j$, $\partial \pi_{Pk}/\partial e_k = 0 \ \forall k$. Thus, conditions (1) and (2) differ from the profit-maximizing conditions. Essentially, the market costs of inputs that increase polluting runoff (i.e. inputs for which $\partial r_i/\partial x_{ij} \geq 0$) are less than the social costs while the private benefits of inputs that reduce polluting runoff (i.e. inputs for which $\partial r_i/\partial x_{ij} \leq 0$) fail to reflect the social value of their use in protecting the environment. Accordingly, farmers will over- (under-) utilize inputs that increase (mitigate) runoff without policies that encourage them to do otherwise. Similarly, point sources will generate emissions in socially excessive amounts.

Conditions (1) and (2) clearly indicate that a coordinated approach involving both point and non-point sources is needed to achieve the least-cost solution. The theory of environmental policy has largely addressed the design of instruments that can satisfy condition (2) (Baumol and Oates, 1988). In the remainder of this section, it is implicitly assumed that point source instruments satisfy condition (2) so that the focus may be on the design of non-point instruments. Subsequent sections present instruments (ambient taxes and trading) that directly coordinate point and non-point pollution reductions.

Griffin and Bromley considered four types of environmental instrument for non-point sources. The instruments they considered are formed by combining two compliance measures (inputs and estimated runoff) with two types of stimulus (taxes and standards), where estimated runoff is simply the non-point production function. Given a farm-specific tax-based estimated runoff, farms will maximize after-tax profits, $\pi_{Ni}(x_i) - t_i r_i$, by equating the marginal gain in pre-tax profits from the use of the input with the marginal cost of the increased tax payment, i.e. $\partial \pi_{Ni} / \partial x_{ij} = t_i \partial r_i / \partial x_{ij} \ \forall i, j$. Comparing this marginal condition with equation (1), it is clear that tax rates of the form $t_i = \lambda^*(\partial W^*/\partial a)(\partial a^*/\partial r_i) \ \forall i$ can be used to obtain the cost-effective outcome. If $W(a) = D(a)$, then the optimal tax rate on estimated emission is the Pigouvian rate. Alternatively, for an estimated runoff standard, farms maximize profit subject to estimated runoff being restricted at or below a target level. The least-cost solution is achieved provided that the targets are of the form $r_i^* = r_i (x_i^*) \ \forall i$.

Given input-based tax instruments, farms will maximize after-tax profits, $\pi_{Ni} (x_i) - \sum_{j=1}^{m} \tau_{ij}$, by equating marginal gain in pre-tax profits from the use of each input with the corresponding tax input rates, $\partial \pi_{Ni} / \partial x_{ij} = \tau_{ij} \ \forall i, j$. Comparing this marginal condition with equation (1), it is clear that tax rates of the form $\tau_{ij} = \lambda^*(\partial W^*/\partial a)(\partial a^*/\partial r_i)(\partial r_i^*/\partial x_{ij}) \ \forall i, j$ obtain the cost-effective solution. Alternatively, the cost-effective allocation may be attained using farm-specific input standards that limit the use of pollution-increasing inputs to no more than their optimal levels and require the use of pollution-control inputs at no less than their optimal levels. Thus, Griffin and Bromley showed that the choice of base is not all that important as long as the instrument is set at an appropriate level for the chosen base.

Stochastic emissions and imperfect information about fate and transport

Griffin and Bromley's model enlarged the domain of economic thinking on non-point instruments with theoretical support to include emissions proxies and input-based instruments, and provided a significant foundation for subsequent theoretical and empirical work. However, their model missed two important features of the non-point problem. Firstly, non-point emissions in their model are a deterministic function of farmers' choices of inputs – there are no inherently stochastic components. Secondly, they assumed that the non-point production function is known. These assumptions also apply to the fate and transport function a.[8] Under these assumptions, observation of input choices allows a perfect forecast of non-point emissions and their fate. Accordingly, observed input choices are a perfect substitute for metering non-point emissions. These assumptions are limiting. Nonpoint emissions are inherently stochastic, principally because stochastic weather variables (precipitation, wind, temperature, etc.) are key forces behind the formation, transport and ultimate fate of pollutants from agricultural land (Shortle and Dunn, 1986; Segerson, 1988). A more general version of the non-point production function will therefore include stochastic weather variables. Further, knowledge of the relations governing the formation, fate and transport of agricultural pollutants is imperfect. Under these conditions, observations of inputs cannot give perfect forecasts of non-point emissions or their environmental impacts.

Shortle and Dunn (1986) presented a model in which non-point emissions are unobservable and stochastic, fate and transport processes are stochastic, and there is imperfect information about these relations.[9] In terms of non-point emissions, these features are incorporated into the model by adding stochastic environmental variables (e.g. weather), v_i, and random variables related to imperfect information about model specification, θ_i, into the ith farm's non-point production function, i.e. $r_i = r_i(x_i, \alpha_i, v_i, \theta_i)$, where x_i and α_i are as defined above. Under this specification, plugging observations of farm inputs into a non-point production function is no longer a perfect substitute for measuring runoff without error. Accordingly, farms cannot control their actual runoff with certainty. Instead, they can make production and pollution control choices to influence the distribution (or probabilities) of possible runoff levels. Environmental outcomes are also stochastic in this framework since runoff is stochastic. Moreover, environmental processes such as the fate and transport of pollutants may be stochastic in their own right. Denote ambient pollution concentrations by $a = a(r_1, ..., r_n, e_1, ..., e_s, \zeta, \psi, \delta, \theta_a)$, where ζ and ψ are as defined above, δ represents stochastic environmental variables that influence transport and fate, and θ_a represents random variables related to imperfect information about transport and fate.

With stochastic environmental outcomes, useful notions of cost-effectiveness for agricultural non-point pollution control must consider variations in the ambient impacts of different sources and the natural variability of non-point loadings. There are several possibilities. The simplest is a combination of point and non-point pollution control efforts that minimizes costs subject to an upper bound on the expected ambient concentration. For instance, if the ambient target is a_0, then the proposed allocation minimizes costs (maximizes expected profits) subject to $E\{a(\cdot)\} \le a_0$. However, this allocation may not have economically desirable properties. An important limitation of this constraint is that it does not explicitly constrain the variation in ambient pollution. An allocation that satisfies the constraint could in principle result in frequent harmful violations of the target. Another approach to defining least-cost allocations uses probabilistic constraints of the form Prob $(a \ge a_0) = \Phi(a) \le \alpha$ $(0 < \alpha < 1)$. This 'safety-first' approach has received attention in economic research on pollution control when ambient concentrations are stochastic and is consistent with regulatory approaches to drinking water quality and other types of environmental protection (Beavis and Walker, 1983; Lichtenberg and Zilberman, 1988; Lichtenberg *et al.*, 1989; Shortle *et al.*, 1999).

The effect of the constraint is illustrated in Fig. 2.1. The ambient impacts of farms' profit-maximizing production decisions in the absence of government intervention are illustrated by the probability density function (pdf) f_π. The probability of ambient pollution levels exceeding a_0 is given by the area under the tail of the distribution to the right of a_0. This probability is well in excess of α. The impact of environmental policies to achieve the safety-first

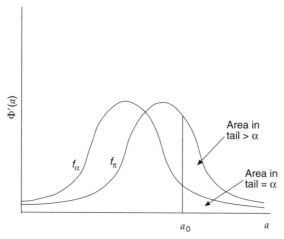

Fig. 2.1. The 'safety-first' probability constraint P $(a \leq a_0) \leq \alpha$.

constraint is to shift the pdf to the left so that the area in the tail is less than or equal to α. Such a case is illustrated by the pdf f_α.

The safety-first constraint will be satisfied as an equality for the cost-effective outcome. However, there are many pdfs for which the probability of exceeding a_0 equals α. The multiplicity of eligible pdfs is illustrated in Fig. 2.2, where two are presented, f_{nc} and f_c, that both satisfy the constraint. Density f_c has a smaller mean ambient concentration but greater variance than density f_{nc}. The economic problem is to determine which pdf is the most cost-effective. Suppose that f_c is the pdf realized in a cost-effective approach. In this case, farmers would find it less costly to adopt measures that reduce the mean ambient pollution than measures that satisfy the constraint by reducing the variance of ambient pollution. The reverse would be true if f_{nc} were the pdf realized in the least-cost solution.

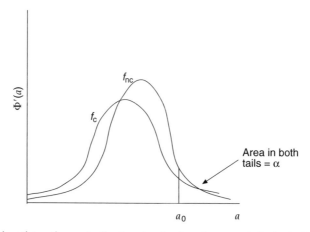

Fig. 2.2. Identifying the cost-effective density function to satisfy the 'safety-first' probability constraint P $(a \leq a_0) \leq \alpha$.

This comparison illustrates an important aspect of the safety-first approach. With a smaller variance, the distribution f_{nc} may result in smaller expected damages than those resulting under the distribution f_c. If the difference in expected damages is large enough, then the distribution f_{nc} may actually be more economically efficient than f_c in satisfying the constraint (Shortle, 1990; Horan, forthcoming).

The economically most interesting cost-effectiveness concept when a is stochastic is an upper bound on expected damage costs. Only in this case will allocations that achieve the target at least cost be unambiguously more efficient than allocations that achieve the target at higher cost (Shortle, 1990; Horan, forthcoming).[10] Thus, a more cost-effective outcome is not necessarily more efficient when damage costs are not a consideration.

Consider an environmental target of the general form:

$$E\{W(a)\} \le T \tag{3}$$

where $W(a)$ is again continuous and increasing. Constraint (3) is an expected damage cost constraint if $W(a) = D(a)$. For heuristic purposes, $W(a)$ is taken to be the damage cost function; however, the relation defined by $W(a)$ is fairly general and could just as easily encompass a variety of other types of environmental quality constraint. For example, the constraint is simply an upper bound on the expected ambient concentration if $W' = 1$. If $W(a)$ is quadratic, then the constraint implies an upper bound on a linear combination of the mean and variance of the ambient concentration (Samuelson, 1970), which could represent a deterministic equivalent of probabilistic constraints for some pdfs (Vajda, 1972). $W(a)$ could also represent $\Phi(a)$, in which case the expectations operator vanishes.

The cost-effective or first-best allocation associated with constraint (3) solves:

$$\max_{x_{ij}, e_k} NB = \sum_{i=1}^{n} \pi_{Ni}(x_i) + \sum_{k=1}^{s} \pi_{Pk}(e_k),$$

subject to (3). With appropriate continuity and convexity assumptions, first order necessary conditions for the cost-effective plan are:

$$\frac{\partial \pi_{Ni}}{\partial x_{ij}} = \lambda E\left\{ W^9(a) \frac{\partial a}{\partial r_i} \frac{\partial r_i}{\partial x_{ij}} \right\} \forall i,j \tag{4}$$

$$\frac{\partial \pi_{Pk}}{\partial e_k} = \lambda E\left\{ W^9(a) \frac{\partial a}{\partial e_k} \right\} \forall k \tag{5}$$

where λ is the shadow value of (3). Conditions (4) and (5) are analogous to conditions (1) and (2), except that they equate marginal profits with *expected* marginal damages instead of *actual* marginal damages. Denote the solution to equations (4) and (5) by x_{ij}^* $\forall i,j$, e_k^* $\forall k$, and λ^*.

We again look at non-point policy design assuming that condition (4) is satisfied by the choice of a point source instrument. Whereas Griffin and Bromley (1983) found the choice of base is unimportant in a deterministic model, Shortle and Dunn (1986) demonstrated that the choice of base is important when runoff is stochastic. Specifically, they found that instruments based on estimated non-point production functions (estimated runoff) only provide farmers with incentives to consider how their choices impact estimated or mean runoff and not other moments of runoff (such as the variance) that could have important impacts on expected damages.

To illustrate, consider a tax policy based on runoff estimates. Denote t_i as a farm-specific tax rate applied to the estimated runoff of the *i*th site. The producer's after-tax profits are given by $\pi_i(x_i) - t_i E\{r_i\}$.[11] Because the tax is based on estimated runoff, the government must provide farmers with information regarding the relationship $E\{r_i\}$ (where E represents the government's expectations operator) so that farmers understand how their choices influence their tax. Such information could be provided by access to a computer model or simply from a schedule that indicates different levels of runoff that are expected to result from various management choices. Given the tax, farmers' first order necessary conditions for input use are, for an interior solution,

$$\frac{\partial \pi_i}{\partial x_{ij}} - t_i E\left\{\frac{\partial r_i}{\partial x_{ij}}\right\} = 0 \quad \forall i,j \tag{6}$$

Comparison of condition (6) with condition (4) implies that the following condition must hold to obtain the cost-effective solution:

$$t_i = \frac{\lambda^* E\left\{W'(a^*)\frac{\partial a^*}{\partial r_i}\frac{\partial r_i^*}{\partial x_{ij}}\right\}}{E\left\{\frac{\partial r_i^*}{\partial x_{ij}}\right\}}$$

$$= \lambda^* E\left\{W'(a^*)\frac{\partial a^*}{\partial r_i}\right\} + \frac{\lambda^* cov\left\{W'(a^*)\frac{\partial a^*}{\partial r_i},\frac{\partial r_i^*}{\partial x_{ij}}\right\}}{E\left\{\frac{\partial r_i^*}{\partial x_{ij}}\right\}} \quad \forall i,j \tag{7}$$

The optimal tax rate would therefore be site-specific, to reflect each farm's individual impact on ambient pollution, and equal to the expected marginal damages from runoff plus a covariance term that acts as a risk-premium or reward, depending on the sign.

In general, equation (7) is overdetermined as the single tax rate t_i is determined by m equations. If $m > 1$, there will not, in general, exist a t_i that will satisfy all m equations in (7) and so there is no tax rate that can produce the cost-effective outcome. The intuition here is straightforward. An estimated runoff tax as in equation (7) only provides producers with an incentive to choose input levels to control mean runoff levels. However, choices that cost-effectively reduce $E\{r_i\}$ do not necessarily reduce $E\{W\}$ cost-effectively when W is non-linear. This is because, in addition to mean runoff levels, $E\{W\}$ depends on the variance and other moments of the distribution of r_i and also moments of a, and these other moments will generally depend on each of the m actions taken to reduce mean runoff (Shortle and Dunn, 1986; Shortle, 1990). For example, if the actions taken to reduce $E\{r_i\}$ also have the effect of increasing $var\{r_i\}$ or $var\{a\}$, then $E\{W\}$ may increase. Thus, estimated runoff-based incentives cannot generally be cost-effective (analogous results arise for the cases of regulations and market-based approaches, i.e. trading).

Only when a single input influences runoff (i.e. $m = 1$) or when the covariance between marginal damages and marginal runoff, $cov\{\partial W'(a^*)\partial a^*/\partial r_i, \partial r_i^*/\partial x_{ij}\}$, is zero for each input for each farm (e.g. when $W = \Sigma\, E\{r_i\}$) do the choices that cost-effectively reduce mean runoff also cost-effectively reduce expected damages (Horan et al., 1998). However, it is unrealistic to believe that either of these conditions will generally be satisfied, unless the policy goal is a linear transformation of expected runoff.

Now consider an input-based approach. Obviously, input-based regulations set at the levels x_{ij}^* $\forall i,j$ can be used to provide the cost-effective outcome. Incentives can also be used. Consider a system of farm-specific, per unit input taxes. Denote the tax rate on input j of farm i by τ_{ij} so that expected after-tax profit is $\pi_i(x_i) - \sum_{j=1}^{m} \tau_{ij} x_{ij}$. Note that, in contrast to the case of estimated runoff taxes, farmers only require information on their own input choices to evaluate their tax. Given that producers choose inputs to maximize after-tax profits, the optimal marginal input tax rates (after some manipulation) are:

$$\tau_{ij} = \lambda^* E\{W'(a^*)\} E\left\{\frac{\partial a^*}{\partial r_i}\right\} E\left\{\frac{\partial r_i^*}{\partial x_{ij}}\right\} + \lambda^* E\{W'(a^*)\} cov\left\{\frac{\partial a^*}{\partial r_i}, \frac{\partial r_i^*}{\partial x_{ij}}\right\}$$
$$+ \lambda^* cov\left\{W'(a^*), \frac{\partial a^*}{\partial r_i}\frac{\partial r_i^*}{\partial x_{ij}}\right\} \quad \forall i,j \tag{8}$$

The optimal tax rate for input j for farm i equals expected marginal damages, multiplied by the expected marginal increase in ambient pollution levels from farm i's emissions, multiplied by the expected increase in runoff from increased use of input j at the margin, plus two covariance terms that act as risk premiums or rewards, depending on the signs.

The tax rate may be positive or negative, depending on the signs and relative magnitudes of the three RHS terms in equation (8). The sign of the first RHS term will be positive for pollution-increasing inputs and negative for pollution-decreasing inputs. The signs of the risk terms are ambiguous without further specification. If a is convex (concave) in emissions, then the first covariance term is of the same (opposite) sign as $\partial var(r_i^*)/\partial x_{ij}$. Thus, when a is convex (concave), risk and hence τ_{ij} are increased (decreased) when an increase in the use of the input increases the variance of emissions.[12] Similarly, when $W''' > 0$ ($W''' < 0$), risk and hence τ_{ij} are increased (decreased) when an increase in the use of the input increases the variance of a.

Entry and exit

We have this far described instruments that can lead to optimal input decisions by agricultural producers. The focus has been on changing how producers produce. Our analysis has taken the set of producers as given, but this is limiting in the long run. Firstly, since the instruments we have described will affect profits, they should also affect decisions about whether and where to produce, as well as how to produce. Secondly, least-cost achievement of environmental objectives requires not only that those who are producing in given locations make optimal choices, but also that the number and location of producers is optimal with respect to the standard (Baumol and Oates, 1988). The incentive and regulatory systems we have described may require additional elements to achieve optimal results at the extensive-margin as well as at the intensive-margin agricultural production.

Without loss of generality, larger values of i and k are assumed to correspond to polluters that are socially less efficient, i.e. that either are less productive or, in the case of non-point sources, have locations that are more conducive to the creation of non-point emissions (Horan *et al.*, 1998). This specification allows us to address entry and exit by choosing the values of n and s (i.e. the marginal polluters from each source category) optimally. Specifically, the following two conditions define the cost-effective allocation along with conditions (4) and (5):

$$\frac{\Delta NB}{\Delta n} \approx \pi_{Nn} - \lambda E\{\Delta_n W(a)\} \approx 0 \tag{9}$$

$$\frac{\Delta NB}{\Delta s} \approx \pi_{Ps} - \lambda E\{\Delta_s W(a)\} \approx 0 \tag{10}$$

where $\lambda E\{\Delta_n W(a)\} = \lambda E\{W(a(r_1, ..., r_n, \cdot)) - W(a(r_1, ..., r_{n-1}, \cdot))\}$ is the difference in expected damages from when production occurs on the nth site when the nth site is taken out of production – i.e. the incremental effect of site n on expected external costs, where external costs are defined in terms of the constraint (2) – and $\lambda E\{\Delta_s W(a)\} = \lambda E\{W(a(e_1, ..., e_s, \cdot)) - W(a(e_1, ..., e_{s-1}, \cdot))\}$ is the difference in expected damages from when the sth firm produces and when it does not (i.e. the incremental effect of firm s on expected external costs). Denote the solution to equations (4), (5), (9) and (10) by $x_{ij}^* \, \forall i,j,n^*,e_k^* \, \forall k,s^*$, and λ^*.

Condition (9) describes the incremental impact of site n on expected net benefits. If the nth site is defined optimally, then the addition of any other site will have a negative incremental impact. Thus, production optimally occurs on sites for which the expected incremental benefits of pollution control (i.e. the difference between expected damages when the site is used for the next best alternative and expected damages when production occurs on the site) are greater than the expected incremental costs of pollution control. In the absence of external costs, conditions (4) and (9) require the marginal farm to operate at minimum average cost, or at the point where there are constant returns to scale. Assuming convex damages ($W'' \geq 0$), all farms should operate where price is greater than the private average cost of production in the optimum, although profits will be larger for infra-marginal farms.[13] Analogous interpretations arise for the point source conditions (5) and (10).

The input non-point instruments that we have considered to this point have been designed only to satisfy condition (4). There is no guarantee that the set of producers will be cost-effective, i.e. that (9) will be satisfied as well. It is therefore necessary to use additional instruments that are designed to influence entry and exit without distorting input choices. One such instrument is a lump-sum tax charged to the extra-marginal producers. Denote the lump-sum tax applied to non-point sources by κ_{Ni} and the lump-sum tax charged to point sources by κ_{Pk}. The lump-sum taxes can be set such that extra-marginal producers will expect to earn negative profits if they produce, i.e.:

$$\kappa_{Ni} > \pi_{Ni}\left(x_i^{**}\right) - \sum_{j=1}^{m} \tau_{ij}^* x_{ij}^{**} \quad \forall i > n^* \tag{11}$$

where the superscript ** denotes the values of variables that would result if the sources did produce (Shortle *et al.*, 1998). The lump-sum taxes (11) ensure that the extra-marginal producers are better off when they do not produce.[14] It is not necessary to impose a lump-sum tax on the marginal or infra-marginal producers. However, a lump-sum subsidy will be necessary for the marginal and/or any infra-marginal producer whose decision to produce is adversely influenced by the magnitude of cost-effective taxes or, in the case of regulations, producers' costs of compliance. While we have not addressed the mechanisms for point source pollution control, similar issues arise and have been addressed in the literature (Baumol and Oates, 1988).

Dynamics

Our analysis of input-based instruments to this point has been static. We have not considered issues such as capital accumulation (particularly investments in abatement capital) in the presence of adjustment costs, the timing of nutrient and chemical applications, nutrient carryover within a field (e.g. phosphorus can build up in soils and be available for many years), intra-annual events that might influence runoff and transport, and pollution accumulation within a body of water. While many of the insights gained from the static analysis are relevant for the dynamic case, additional policy implications can come from a dynamic analysis. The most significant of these relate to how policies optimally evolve over time to ensure cost-effective rates of investment in pollution control equipment and environmental improvements. An inter-temporal analysis of input-based instruments and emissions proxies is beyond the scope of this chapter. Contributions on dynamic aspects include articles by Xepapadeas (1991, 1992, 1994), Kim *et al.* (1993), Dosi and Moretto (1993, 1994), Eiswerth (1993) and Tomasi *et al.* (1994).

Towards the Application of Incentives and Regulations on Inputs, Practices and Emissions Proxies

The theoretical results that we have presented provide rules for the design of cost-effective or *first-best* policies. However, like Pigouvian emissions taxes, the transactions costs associated with identifying and implementing optimally designed instruments would be prohibitive. The instruments are exceptionally information intensive and complex in that all choices that affect environmental outcomes are subject to the incentives or regulations, with the tax/subsidy rates or regulations being farm-specific to reflect the differential environmental impacts of polluting runoff from different locations on ambient environmental conditions. Developing and administering such instruments would not be a trivial problem given the large numbers of agents, technical complexity and hetero-geneous conditions that characterize many agricultural non-point pollution problems. In particular, farm-specific taxes would pose significant enforcement problems because they provide incentives for 'under the table' arbitrage between farms that could defeat the efficiency of the tax system. In practice, uniform rates, at least within regions where such arbitrage could easily occur, would be required. Similarly, while purchases of potentially harmful inputs or investments in pollution control structures can be easily tracked in some instances, many farm management decisions having a large impact on the environment are too costly to monitor or verify. These information costs require the tax/subsidy base to be reduced to a subset of choices that are both relatively easy to observe and correlated with ambient impacts. Instruments that are developed to minimize costs, subject to the environmental constraint (3) *and* subject to the additional restrictions of being suboptimally differentiated across producers and of being applied to suboptimal instrument bases, are classified as *second-best*. Plausible instruments will necessarily be of a second-best type.

Excluding transactions costs, second-best instruments will not be cost-effective relative to the first-best instrument designs. For instance, increased uniformity of the tax/subsidy rates or regulations across polluters will reduce the cost-effectiveness of pollution control because it diminishes the potential gains from differential treatment of polluters according to their relative impacts on ambient conditions. High control cost or low damage cost polluters will end up devoting too many resources to pollution control while low control cost or high damage cost polluters will devote too few resources to pollution control. Similarly, cost-effectiveness is reduced as the set of inputs subject to (or targeted by) incentives or regulations is reduced from the optimum because more intense control will be required for the targeted set to compensate for the absence of incentives for control of the non-targeted set. The determination of which inputs are likely to be the best prospects for targeting of instruments will depend on the nature of any resulting input and other substitution effects, correlation with environmental quality, and enforcement and monitoring costs. If the differences in the cost-effectiveness of first- and second-best designs are small before transactions costs are considered, then even a small saving in transactions costs may be justified. If the differences are large, then the savings in transactions costs must be comparably large.

While theory can help to guide the choice and design of second-best instruments (e.g. Shortle *et al.*, 1998), questions about compliance measures, types of instrument and other details are inherently empirical. Experience provides little for economists to work with in evaluating alternatives since there has been only limited use of input-based economic incentives and regulations for agricultural non-point pollution control. The little information that is available suggests that many existing programmes are not very cost-effective. For instance, Freeman (1994) concluded that costs associated with the US Clean Water Act very likely substantially outweigh the realized benefits. One reason is the very limited focus on non-point sources when these sources are significant contributors to water quality impairments. The USDA Conservation Reserve Program (CRP) is often touted as providing large environmental benefits, particularly since participation in the programme became based on an Environmental Benefits Index. Still, 'most agree that the overall programme could have been structured to provide even greater benefits. In addition, the government cost of enrolling some CRP acres could have been lower, particularly in the Great Plains' (USDA ERS, 1997). Moreover, compared with other approaches, land retirement is considered to be a relatively expensive method of achieving water quality improvements and other environmental benefits (USDA ERS, 1997). Even USDA programmes such as the Water Quality Incentives Program and the Water Quality Program, which are more focused on changing production decisions at the intensive margin, tend to encourage practices that increase net farm returns (USDA ERS, 1997). The implication is that the effectiveness of the programme in bringing about significant water quality improvements is limited because the types of on-farm changes that are needed to create these improvements, which often come at a cost to producers, are not being made.

Given the limited experience with actual instruments, empirical research on the design and performance of instruments to reduce water pollution from agriculture is largely conducted using models that simulate economic and ecological impacts, rather than through *ex post* analysis. Several studies are summarized in Table 2.2. Methods used in these models are discussed in greater detail in Chapter 4.

Table 2.2. Empirical studies of agricultural pollution.

| Study | Study area | Pollutant | Physical model | | | | | Economic model | | | Policy features |
| | | | Scale[a] | Emissions impact heterogeneity[b] | Random variation | | Control cost heterogeneity[c] | Private control cost uncertainty[d] | Static/ dynamic | Instruments assessed[e] | Asymmetric public and private information |
					Emissions	Fate/ transport					
Jacobs and Casler (1979)	Fall Creek Watershed, New York State, USA	Phosphorus	Watershed	No	No	No	Yes	No	Static	Emissions taxes, emissions standards	No
Park and Shabman (1982)	Occoquan River Basin, VA	Phosphorus	Region	No	No	No	No	No	Static	Input standards	No
Milon (1987)	Honey Creek Basin, OH	Pesticides and nutrient	Field	Yes	Yes	No	Yes	No	Static	Input standards	No
Gardner and Young (1988)	Colorado River Basin	Salinity	Watershed	No	No	No	No	No	Static	Input standards	No
Braden et al. (1989)	Long Creek of Macon City, IL	Sediment	Field	Yes	No	No	Yes	No	Static	LCA	No
McSweeny and Shortle (1990)	Nansemond River and Chuckatuck Creek Watershed, VA	Nitrogen	Farm	No	Yes	No	No	No	Static	LCA	No
Braden et al. (1991)	Pipestone Creek, Galien River, MI	Sediment and pesticides	Field	No	Yes	No	No	No	Static	Input standards	No

Continued

Table 2.2 *Continued*

Study	Study area	Pollutant	Physical model Scale[a]	Emissions impact heterogeneity[b]	Random variation Emissions	Fate/ transport	Economic model Control cost heterogeneity[c]	Private control cost uncertainty[d]	Static/ dynamic	Policy features Instruments assessed[e]	Asymmetric public and private information
Johnson et al. (1991)	Columbia Basin, OR	Nitrate	Farm	Yes	Yes	No	Yes	No	Dynamic	Estimated emissions tax/ standards, input tax/ standards	No
Taylor et al. (1992)	Willamette Valley, OR	Nitrogen, phosphorus and sediment	Farm	Yes	No	No	Yes	No	Static	Input tax, estimated emissions standards and input standards	No
Bernardo et al. (1993)	Central High Plain (CO/KS/ NM/OK/TX)	Nitrogen, pesticides	Watershed	NA	Yes	NA	Yes	No	Static	Input standards	No
Lee et al. (1993)	Colorado River Basin (AZ/CO/UT)	Salinity	Watershed	Yes	Yes	No	Yes	No	Static	LCA	No
Mapp et al. (1994)	Central High Plain, USA	Nitrogen	Watershed	NA	Yes	NA	Yes	No	Static	Input standards	No
Moxey and White (1994)	Northern England	Nitrate	Field	Yes	No	No	Yes	No	Static	Emissions standards, input standards	No

Study	Location	Pollutant	Scale						Static/Dynamic	Instrument	
Ribaudo and Bouzaher (1994)	Midwest (Corn Belt)	Atrazine	Region (multi-state)	Yes	Yes	Yes	No	Yes	Static	Emissions standards and input standards	No
Shortle and Laughland (1994)	USA	Nitrogen and pesticides	NA	NA	NA	NA	No	No	Static	Input taxes and input standards	No
Shumway and Chesser (1994)	Crop Reporting District of south central TX	Pesticides	Region	No	No	No	No	No	Static	Input tax	No
Abler and Shortle (1995)	USA	Fertilizer and pesticides	NA	NA	NA	NA	No	No	Static	Green technology development	No
Helfand and House (1995)	Salinas Valley, CA	Nitrates	Field	No	No	No	Yes	No	Static	Input taxes and input standards	No
Teague et al. (1995)	Central High Region (OK)	Nitrates, pesticides	Farm	NA	Yes	NA	No	Yes	Static	LCA	No
Wu and Segerson (1995)	WI	Pesticides, fertilizer	NA	NA	NA	NA	Yes	No	Static	Input taxes, output taxes	No
Flemming and Adams (1996)	Malheur County, OR	Nitrate	County	Yes	No	No	Yes	No	Dynamic	Input taxes	No

Continued

Table 2.2 *Continued*

| Study | Study area | Pollutant | Physical model | | | | | Economic model | | Policy features | |
			Scale[a]	Emissions impact heterogeneity[b]	Random variation Emissions	Random variation Fate/transport	Control cost heterogeneity[c]	Private control cost uncertainty[d]	Static/ dynamic	Instruments assessed[e]	Asymmetric public and private information
Hopkins et al. (1996)	Eastern Corn Belt Plain and Erie Huron Lake Plain, OH	Nitrate-nitrogen; phosphorus; erosion; pesticides	Field	No	No	No	Yes	No	Static	Input taxes; emissions taxes, emissions standards, design standards	No
Huang et al. (1996)	Kanawha, IA	Nitrates	Field	No	No	No	No	Yes	Dynamic	Input standards	No
Larson et al. (1996)	Salinas Valley, CA	Nitrates	Field	No	No	No	Yes	No	Static	Input taxes	No
Rigby et al. (1996)	NW England	Nitrogen, phosphorus	Region	Yes	No	No	Yes	No	Static	Input standards	No
Weinberg and Kling (1996)	San Joaquin Valley, CA	Chemicals in drainage	Watershed	No	No	No	Yes	No	Static	Input tax, emissions tax, marketable permits	No
Gren et al. (1997)	Baltic Sea Drainage Basins	Nitrogen, phosphorus	Watershed	Yes	No	No	Yes	No	Static	LCA	No
Randhir and Lee (1997)	White River Basin, IN	Nitrogen, phosphorus, soil loss, atrazine, alachlor	Watershed	No	No	No	No	Yes	Static	Input taxes and input standards	No

Study	Location	Pollutant	Scale							Dynamics	Instrument	
Vatn *et al.* (1997)	SE Norway	Nitrogen, phosphorus and sediment	Watershed	Yes	No	No	Yes	No	No	Dynamic	Input taxes, input standards	No
Yiridoe and Weersink (1998)	Canada	Nitrogen	Farm	No	No	No	No	No	No	Static	LCA	No
Carpentier *et al.* (1998)	Lower Susquehanna Watershed	Nitrogen	Watershed	No	No	No	Yes	No	No	Static	Performance standard	No
Qiu *et al.* (1998)	Goodwater Creek Watershed, MO	Nitrogen, sediment	Watershed	Yes	Yes	Yes	Yes	Yes	Yes	Static	LCA	No
Horan *et al.* (2001)	Susquehanna River Basin, PA	Nitrogen	Watershed	Yes	Yes	Yes	Yes	Yes	No	Static	LCA, point/non-point trading	No
Claassen and Horan (2001)	North Central USA (IA, MI, MN, WI, MO, OH, IN, IL)	Nutrients	Multi-state	No	No	NA	Yes	No	No	Static	Input taxes	No

Notes: We have done our best to ensure the accuracy of this table. However, because not all studies describe each of the features in much detail, there is a possibility that some of our interpretations may be in error in some instances.

[a] Site parameters and exogenous variables defining the nonpoint emissions function r_i will generally differ by source for watershed models involving multiple sources.

[b] Impact heterogeneity is present if there are explicitly modelled multiple polluters and the marginal environmental impacts of non-point emissions differ by source (i.e. $\partial a/\partial r_i \neq \partial a/\partial r_j, \forall i \neq j$).

[c] Cost heterogeneity is present if there are multiple non-point source units in the study area and these different source units have different costs.

[d] Private control cost uncertainty is unrelated to random variations in emissions, which in a technical sense would lead to uncertainty in the level of pollution control and hence control costs. Instead, we are referring to other types of uncertainty, such as uncertainty in markets or production (e.g. caused by weather or imperfect understanding of technological relationships) that affect economic returns to producers and hence the costs they incur (in terms of reductions in economic returns) from undertaking measures to reduce emissions.

[e] LCA refers to the least cost allocation, which is equivalent to implementing cost-effective input standards.

The theoretical literature indicates that comprehensive studies of agricultural environmental instruments should include stochastic pollution processes, heterogeneity among the environmental impacts of different pollution sources, and asymmetric information between the regulatory agency and polluters about polluters' control costs. Very few empirical studies model more than one of these features (Table 2.2). Heterogeneity is the most common theme while only a few studies consider the impacts of stochastic pollution and asymmetric information. In what follows, we discuss the results of some empirical research that addresses some fundamental questions in constructing second-best policy approaches.

Uniform versus differentiated incentives

Several studies have examined the costs of applying taxes or regulations uniformly rather than differentially across farmers (Table 2.2). In one example, Helfand and House (1995) compared several instruments for reducing nitrate leaching by 20% from two soils used for lettuce production in California's Salinas Valley. Nitrate leaching in their model (simulated using EPIC) was increasing in the amounts of nitrogen fertilizers and irrigation water applied, and was unaffected by any other inputs. Accordingly, the instruments they considered were taxes and restrictions on nitrogen inputs, and taxes and restrictions on irrigation water inputs. Their analysis is especially interesting for its comparison of first-best and second-best specifications of the input-based instruments. Specifically, they compared the costs of achieving the target reduction under several scenarios involving combinations of taxes (or restrictions) on one or both inputs, with the rates (restrictions) being either uniform or differentiated across soils. They found the cost-effectiveness of uniform applications of the instruments to be only slightly less than that of differentiated applications. In contrast, they found that cost-effectiveness might be significantly reduced (relative to the first-best outcome) when only a subset of production decisions was targeted by a uniform policy. It is difficult to determine the extent to which this result was due to the instruments being applied uniformly or due to the instruments being applied to only a subset of choices. The two effects cannot easily be separated out. At a minimum, it would be necessary also to know the costs associated with applying farm-specific instruments to the same subset of choices, and these results were not reported. Thus, all we can say is that uniformity could be more of a factor when the instrument base is suboptimal.

Given the limited amount of heterogeneity in the Helfand and House model (only two soils), their results for the case of uniform instruments applied to each input should not be generalized. Most empirical studies show that, transactions costs aside, highly targeted, information intensive strategies for non-point pollution control policies outperform undifferentiated strategies, often by a substantial margin (e.g. Babcock *et al.*, 1997; Flemming and

Adams, 1997; Carpentier *et al.*, 1998; Claassen and Horan, 2001). This may be especially true when market prices are impacted by producer responses to environmental policy. For instance, optimal uniform tax rates are often found to be a weighted average of optimal differentiated tax rates when prices are fixed (e.g. Helfand and House, 1995; Flemming and Adams, 1997; Shortle *et al.*, 1998), but this is not necessarily the case when market prices are endogenous. Claassen and Horan (2001) examined the design and performance of differentiated and uniform fertilizer taxes applied to US corn producers in the Corn Belt and Lake States (Michigan, Wisconsin, Minnesota, Ohio, Illinois, Indiana, Iowa and Missouri) when the goal is a 20% reduction in polluted runoff from each of four sub-regions. This region accounts for 65% of total US corn production, and so the corn price and land, fertilizer, capital and labour prices were all endogenous to the model. Optimal uniform taxes are found to be greater than any single differentiated tax in this setting. The tax causes producers in each region to reduce production, which increases the equilibrium price of corn. In turn, the larger corn price encourages producers in some sub-regions to increase production and hence fertilizer use. Thus, larger taxes are needed to overcome price effects that undermine the tax in some sub-regions.[15] Moreover, the cost of achieving the environmental goals is greater than would be the case if prices were fixed. Finally, Claassen and Horan found that uniform taxes also affect the distribution of economic impacts across landowners by increasing the divergence in landowner returns relative to the case of differentiated taxes.

Choice among bases

A number of studies support the conclusion that the choice of instrument base can significantly influence the cost-effectiveness of agri-environmental policy (e.g. Weinberg *et al.*, 1993b; Larson *et al.*, 1996; Weinberg and Kling, 1996). For instance, Helfand and House (1995) and Larson *et al.* (1996) examined various input-based instruments to limit nitrate leaching from lettuce production in the Salinas Valley, California. Both studies found that instruments based on irrigation water were more cost-effective than instruments based on nitrogen use, because irrigation water was more highly correlated with nitrate leaching. The implication is that the appropriate instrument base may not be the chemicals or nutrients responsible for pollution, but rather the choices that are most highly correlated with pollution flows.

Although we demonstrated above that first-best input-based instruments will outperform instruments based on estimated runoff, this conclusion need not hold once the comparison is between second-best input instruments and emissions proxies. In the imperfect real world of non-point pollution control, instruments that use emissions proxies may well outperform input-based instruments. One important difference between the two instrument classes is that input-based instruments allow for differential targeting of inputs whereas estimated runoff-based instruments do not. Differential treatment of inputs may provide advantages in terms of a better ability to fine-tune input risk effects (i.e. the impact of different inputs on the variance and other moments of environmental outcomes). However, these advantages are diminished with uniform instruments and when instruments are applied to only a limited number of inputs. If the risk effects associated with the use of inputs are very small, then there should be little difference in the performance of the two instrument classes, given that instruments are applied to all inputs that affect runoff. Given that the risk effects are small, we would expect the relative performance of estimated runoff-based instruments to improve relative to those based on inputs, as inputs that affect runoff are excluded. Conversely, if risk effects are important, input-based instruments may be comparatively advantageous, provided that the set of inputs in the targeted set is not overly restricted.

Another important difference between the two instrument classes, in the case of uniform instruments, is that estimated runoff-based instruments have the advantage of transmitting more site-specific information to producers about their environmental pressures (i.e. mean runoff) relative to input-based instruments. The extent of this advantage is likely to depend on the correlation between key environmental and cost relationships. Well-constructed proxies can better correlate with environmental quality impacts than individual inputs when runoff is a function of more than one choice variable. For instance, in the case of nitrogen losses, the residual nitrogen available for leaching into groundwater is more highly correlated with the manageable nitrogen excess than with the fertilizer application (National Research Council, 1993). Application of this approach requires the existence of proxies that are good indicators of measures of environmental pressures and not unduly burdensome to compute and enforce (Braden *et al.*, 1991; Braden and Segerson, 1993). The existence of such measures will vary. In The Netherlands, livestock producers will be charged a fee based on surplus phosphate from manure (Weersink *et al.*, 1998). The nutrient accounting system used to determine surplus phosphate is described in Breembroek *et al.* (1996).

Comparisons involving inputs and estimated runoff or other performance bases are limited and more work is needed in this area. Several studies show that taxes or standards applied to nitrogen emissions proxies (e.g. excess nitrogen or expected nitrate leachate) are more cost-effective than taxes or standards applied to nitrogen inputs (McSweeny and Shortle, 1990; Johnson *et al.*, 1991; Shortle *et al.*, 1992; Fontein *et al.*, 1994; Huang and LeBlanc, 1994). However, these studies have not considered the input-related risk effects.

Asymmetric information and choices among instruments

Our development of the first-best instruments was cognizant of the uncertainty that environmental decision-makers have about biophysical relationships between producers' choices and environmental outcomes, but we assumed that they have perfect information about producers' costs of pollution control activities. In general, however, differences in information sets will be the norm. Because individual farmers have limited expertise and resources for developing alternative technologies that can help farmers to reduce the environmental impacts of production, there is an important role for publicly sponsored research to foster 'green' technology development, and education and technical assistance programmes to disseminate technical information (Chapter 3). Still, economists generally expect that farmers will have specialized private knowledge (which they are not willing to share with regulatory agencies) about the options for pollution control and corresponding costs relevant to their particular situations.

Input tax/subsidy schemes and contractual arrangements that can elicit farmers' specialized knowledge have been described in the literature (Shortle and Abler, 1994; Smith, 1995b; Smith and Tomasi, 1995; Romstadt, 1997). These measures are considerably more complex than the first-best instruments described above because they involve transfers of cost information from individuals to the pollution control authority and other features to encourage individuals to be truthful. These instruments seem to us to be mainly of theoretical interest. Accordingly, plausible water quality programmes must be designed with uncertainty about control costs in addition to the uncertainty about biophysical relationships between producers' choices and environmental outcomes. One consequence is particularly noteworthy. With imperfect information about control costs, producer responses cannot be predicted accurately; therefore, it will be impossible to design tax/subsidy incentives that will exactly satisfy the environmental goal. Actual environmental performance may be better or worse than the target level (Baumol and Oates, 1988).

An important topic, given imperfect and asymmetric information about pollution control costs, and implying uncertainty about how polluters will respond to economic incentives, is the choice between tax/subsidy schemes and quantity controls. With symmetric information about pollution control costs, firm specific input and emissions commands can be issued to achieve the first-best solution. The commands would satisfy conditions (4) and (5). In addition, lump-sum charges or commands would be required to satisfy the entry–exit conditions (9) and (10). Alternatively, tradeable permit systems, which we describe in a subsequent section, could be used. These entail quantity controls in that they restrict the aggregate levels of emissions or inputs. A fundamental result of the literature on instrument choice is that instruments that perform equally under symmetric control cost information need not do so under asymmetric control cost information.

The seminal work is Weitzman (1974), based on a highly simplified model in which emissions are non-stochastic and the ambient concentration is linear in emissions. Contrary to conventional thinking at the time, Weitzman showed that the choice of instrument does matter when firms hold private abatement cost information. Using second-order approximations for the costs and benefits of pollution control, he found that the optimal instrument choice depends on the relative slopes of the marginal benefit and cost functions. Unfortunately, this 'rule of thumb' seems to have been adopted wholesale by economists when in fact it does not appear to be robust. Malcomson (1978) showed that Weitzman's results for relative slopes relied heavily on assumptions about the form of uncertainty and that the opposite results arise for different uncertainty specifications. Stavins (1996) found the results for relative slopes to be overly simplistic when there is correlated uncertainty associated with both costs and benefits. Given these concerns and given the additional complexities inherent in the design of first- and second-best non-point instruments, the temptation to generalize Weitzman's result to choices between incentives and quantity controls on polluting inputs or emissions proxies should be resisted.

Mixed systems of input-based instruments can outperform single instruments when asymmetric information exists. Shortle and Abler (1994) adapted the model of Roberts and Spence (1976) to develop a mixed scheme of tradeable input use permits, subsidies for reduced input use and taxes on input use that would dominate any of the individual elements. Mixed instruments may also perform well even when there is no asymmetric information. For example, Braden and Segerson (1993) discussed mixing an input tax with liability rules.

There is very little empirical literature on the performance of alternative agricultural pollution control instruments under conditions of public uncertainty about polluters' control costs. One exception is Abrahams and Shortle (2000), who compared several tax and standard policies for reducing nitrate pollution in the United States using a model that captures public uncertainty about the costs and benefits of nitrate pollution control. They also computed the value of information that would accrue from resolving the uncertainty about key economic and environmental parameters and the sensitivity of the instruments policy performance to market distortions created by agricultural price and income policies.[16] First-best nitrate policy choices were found to be sensitive to commodity programmes and uncertainty. In particular, they found tax policies to be more cost-effective than quantity controls in the presence of public uncertainty, but with perfect information the quantity instruments are as cost-effective as the tax instruments.

When are regulations generally preferred?

The heavy focus of this chapter on incentives and the empirical findings we have reported indicating that incentives outperform standards in some settings (Helfand and House, 1995; Abrahams and Shortle, 2000) may lead to an incorrect conclusion that regulations are always inferior to incentives. There are instances in which standards may be preferred. Moreover, some cases can be identified in which regulations always make sense (Shortle and Abler, 1997). One notable case is when the expected societal costs of the use of an input or process exceeds the expected benefits for essentially any level of use. Examples are extremely hazardous pesticides, especially when there are close substitutes with lesser risks, and the use of certain types of pesticide in environmentally sensitive areas such as groundwater recharge zones (Weersink *et al.*, 1998). A second case is when techniques exist that have the potential to yield significant environmental gains with little or no cost to the user. For example, several recent studies suggest that nitrogen soil testing in corn production in humid regions of North America can greatly reduce nitrate losses to ground and water resources with little negative economic impact on farmers (US Office of Technology Assessment, 1995). Similarly, no-till farming practices are very effective in reducing sediment pollution problems and can increase farm profits relative to conventional practices in some regions of North America (Eiswerth, 1993; Logan, 1993). Finally, some least-cost techniques are essentially common knowledge. An example is the use of septic systems for domestic wastewater treatment in rural areas.

Some additional practical guidelines for the choice between quantity controls and economic incentives can be based on considerations of the acceptable degree of uncertainty about the level and geographical location of the use of inputs. Compared with fiscal incentives, standards and, to a lesser degree, tradeable permits (described below) have a relatively certain impact on the use of polluting inputs and activities provided that they are adequately enforced. This suggests that quantity controls will be preferable to tax/subsidy schemes, at least as the main mechanism of control, when a high degree of certainty over the level and geographical distribution of polluting inputs and activities is desirable. Included would be pollutants for which the damages are highly uncertain but potentially large and/or irreversible, such as hazardous and toxic substances. Conversely, incentives may be considered advantageous where some uncertainty about emissions is acceptable and where some growth in pollution is considered an acceptable cost of economic growth.

Ambient Pollution Taxes and Liability Rules

Segerson (1988) took a very different approach to the non-point problem than Griffin and Bromley (1982) and Shortle and Dunn (1986). Rather than monitoring the input choices of farmers who are suspected of contributing to environmental degradation, she proposed the use of an ambient-based tax that shifts monitoring from the source of emissions to the receptor. Segerson devised an ambient tax/subsidy scheme that could achieve the efficient outcome for non-point sources as a Nash equilibrium. The scheme pays firm-specific subsidies when the ambient pollution concentration falls below a target and charges firm-specific taxes when the ambient concentration exceeds the target.[17]

For simplicity and without loss of generality, consider the tax portion of this ambient-based incentive. A farm-specific ambient tax can be defined by $t_i a + k_i$, where t_i is the ambient tax rate for farm i and k_i denotes a farm-specific lump-sum tax or subsidy. Each farm will choose input use to maximize expected after-tax profit, $V_i = \pi_i(x_i) - E\{t_i a\} - k_i$. The first-order necessary conditions for an interior solution are

$$\frac{\partial \pi_i}{\partial x_{ij}} - E\left\{ t_i \frac{\partial a}{\partial r_i} \frac{\partial r_i}{\partial x_{ij}} \right\} = 0 \quad \forall i,j \tag{12}$$

Horan *et al.* (1998) found that the same problems that limit the cost-effectiveness of estimated runoff-based incentives limit the cost-effectiveness of ambient-based incentives – namely, an optimal tax rate t_i will exist only when $m = 1$ or when the covariance between marginal damages and marginal ambient pollution, $cov\{\lambda W'(a^*),\ \partial a^*/\partial r_i \partial r_i^*/\partial x_{ij}\}$, equals zero for all farms and for all inputs (e.g. as when $W = E\{a\}$, which, as described above, neglects variations in ambient pollution, which could be important). The intuition is the same as above. An ambient tax only provides firms with an incentive to choose input levels to control expected ambient levels. However, reductions in $E\{a\}$ do not necessarily correspond to reductions in $E\{W\}$ when W is non-linear, because the variance and other distributional moments of a may be influenced by the actions that farms take to reduce $E\{a\}$ (Shortle, 1990). If $var\{a\}$ increases as $E\{a\}$ is reduced, for example, then $E\{W\}$ may increase. When only a single input influences runoff, the linear tax scheme optimally manages such risk effects.

Horan *et al.* (1998) (see also Hansen, 1998) identified two cost-effective ambient taxes that apply under less restrictive conditions. The first is the following linear and state-dependent (i.e. it is determined after the realization of all random variables) form:

$$t_i = \lambda^* W'\left(a^*\right) \quad \forall i \tag{13}$$

The tax rate is the marginal damage given the first-best choices. It is conditional on the realization of all random variables. This tax is applied uniformly to all farms. *Ex ante*, the expected tax faced by the farm is $E\{ta\} = E\{\lambda^* W'(a^*)a\}$.

Alternatively, a first-best ambient tax could be defined as a non-linear function of ambient levels, $T_i(a)$, where (see also Hansen, 1998)

$$T_i\left(a\right) = \lambda^* W\left(a\right) \tag{14}$$

Therefore, in contrast to the linear ambient tax in equation (13), each farm pays an amount equal to *total* damages. As with the linear tax, the non-linear tax is state-dependent and applied uniformly across all farms.[18]

Entry and exit considerations are especially important with ambient taxes because ambient pollution, and hence the incentives generated by the tax, depend on the performance of all farms that contribute to the ambient pollution level. The ambient taxes defined by equations (13) and (14) do not ensure long-run efficiency, and so lump-sum taxes or subsidies, analogous to those derived for the input taxes, are required in addition to the ambient tax (see Horan *et al.*, 1998, for a discussion of entry and exit considerations).

The ambient tax schemes presented above have, at first glance, substantial appeal compared with input tax schemes. Firstly, except for the lump-sum charges to induce optimal entry and exit, there is no need to devise farm-specific policies. Secondly, while we have not included point sources in our analysis, if point sources were also present in the region, an ambient tax would optimally coordinate control of point and agricultural sources without the need to develop and implement separate instruments for each type of source.

However, there are three important characteristics that may severely limit what ambient taxes can accomplish in practice. Firstly, producers are not rewarded or penalized according to their own performance. Rather, rewards and penalties depend on group performance. This makes polluters' expectations (or conjectural variations) about the behaviour of other polluters, and the regulator's knowledge of these expectations, critical to the actual design and performance of ambient-based instruments.

A second important feature is that ambient taxes shift the burden of information from regulators to producers. A producer's response to an ambient tax will depend on its own expectations about the impact of its choices, the choices of others and natural events on ambient conditions. In other words, it will depend on the polluter's theory of fate and transportation. However, given that the typical producer has limited technical information and capacity in this area, it is not obvious that this shift makes good economic sense. Cabe and Herriges (1992) and Horan *et al.* (1999b) explore the consequences of asymmetric prior information about transport and fate. If polluters' prior information about transport and fate differs from that of the regulator, then incentives designed on the assumption that they are the same will not have the desired properties. For instance, a polluter may perceive no impact of its choices on pollution. In this case, an ambient tax may have either no impact, or it will result in a decision to escape the tax entirely by ending the suspect activity. The regulator has a choice between adjusting the incentives for the mismatch, educating the polluter, or a combination of both. The data collection and programme design issues involved in systematically measuring mismatches, developing educational programmes that would effectively close them and adjusting incentives for their effects could be enormous. Similar concerns would apply to expectations about the behaviour of other polluters.

Thirdly, ambient incentives cannot produce a first-best outcome by themselves when polluters are risk-averse (Horan *et al.*, 1999b). The reason is that the tax cannot simultaneously manage environmental risk and the risk to producers that is created by the inherent randomness of the tax. Cost-effectiveness can be achieved in this situation only if the ambient tax is used in conjunction with input taxes to manage the additional risk.

Several other considerations deserve mention. Firstly, monitoring ambient conditions can be highly costly and subject to considerable error. This is illustrated by the uncertainty that exists about groundwater quality in many areas. Secondly, changes in observed conditions may have little relationship to contemporary actions. For instance, nitrates and pesticides may take years to move from fields to wells. Accordingly, incentives based on contemporary changes in ambient conditions may bear little relationship to current behaviour. In such cases, the incentives may do little to encourage improved performance. Finally, there is a capricious aspect to ambient-based taxes that would likely limit political and ethical acceptability. In particular, individuals who take costly actions to improve their environmental performance could find themselves subject to larger rather than smaller penalties, due to environmental shirking on the part of others, natural variations in pollution contributions from natural sources, or stochastic variations in weather. Conversely, individuals who behave badly may end up being rewarded by the good actions of their neighbours or nature.

These and other considerations led Weersink *et al.* (1998) to suggest that ambient taxes may be best suited to managing environmental problems in small watersheds in which agriculture is the only source, farms are relatively homogeneous, water quality is readily monitored and there are short time lags between polluting activities and their water quality impacts. We would add, especially until there is significant real or experimental evidence on how individuals respond to ambient incentives, that the scope of environmental problems should be limited. The theory has been developed for a single pollutant, and asks for fairly complex decisions in response to expectations about the pollutant. With multiple pollutants, particularly if they interact, the decision environment would be much more complex.

Liability rules

Liability represents another form of ambient-based incentive. Individuals who are damaged monetarily or otherwise by the activities of others may have the right to sue for damages in a court of law. If the suit is successful, the court may be guided in its compensation decision by a rule of law or precedent, known as a liability rule. Two important classes of liability rules are: (i) strict liability; and (ii) negligence. Under strict liability, polluters are held liable for full payment of any damages that occur. Under a negligence rule, polluters are only liable if they failed to act with the 'due standard of care' (Segerson, 1995). For example, a producer may not be found negligent (and hence liable) in contaminating groundwater with pesticides if the pesticide was applied in accordance with the manufacturer's specification and the laws regarding application procedures.

The liability rule, while imposed *ex post*, serves as an *ex ante* incentive to deter producers from engaging in activities that may be damaging to others. If polluters feel that their production decisions may result in damages for which they may be held liable, then they must weigh the benefits from participating in pollution-related activities against any penalties they may expect to face as a result of their actions. Thus, liability rules are a form of ambient-based incentive in that they are imposed after ambient pollution and damages are realized (Shavell, 1987). However, liability rules differ from traditional ambient-based incentives because they are only imposed if a suit is privately or publicly initiated, and if a court of law rules in favour of the damaged parties.[19] Instances may therefore arise in which damages occur but no payments are made. Thus, unlike incentives, liability rules do not use markets to create or alter prices (Segerson, 1995).

In the case of strict liability, suppose that the extent of a producer's liability depends on the damages that arise as a result of the ambient pollution level. This liability can be represented by a liability rule, $L_i(a)$.[20] Producers are only held liable if they are sued by a damaged third party and are found to be responsible. Therefore, in addition to the uncertainty they face about a, producers face uncertainty about whether or not they will be sued and held liable. Producers have their own beliefs regarding the probability that they will be sued and held liable, denoted by $q_i(a, \eta_i)$, where η_i is a vector of random variables that may influence this probability. Thus, the ith producer's expected liability payment is $E_i\{q_i(a, \eta_i)L_i(a)\}$. It is easy to see that the liability rule is very similar to a (non-linear) ambient tax.

Turning to negligence, there has been a movement at both the US state and federal levels to hold producers liable for damages resulting from chemical use only if they failed to apply registered chemicals in accordance with the manufacturer's instructions and any related laws (Wetzstein and Centner, 1992; Segerson, 1990, 1995). For example, the Comprehensive Environmental Response, Compensation and Liability Act (CERCLA) restricts producers' liability in this manner. A substantial number of states make compliance with acceptable agricultural best management practices a defence to nuisance actions (ELI, 1997). Negligence rules of this sort are in accordance with the philosophy that producers have a basic 'right to farm' and that they should not be penalized as long as they adhere to standard, accepted practices.

Under a negligence rule, producers are only held liable if they failed to use the 'due standard of care'. The standard can be defined in terms of either damages or input use. That is, liability only applies if a particular level of damages is exceeded or if the use of polluting inputs is excessive. A formal analysis of negligence rules is beyond the scope of this chapter, but an important difference from strict liability is that the lump sum components of negligence rules cannot generally ensure optimal entry and exit *and* that the aggregate liability equals total damages. Instead, negligence rules will not necessarily be effective in limiting entry, because polluters may all avoid liability by producing at suboptimal levels (Miceli and Segerson, 1991) and production may be profitable on more than the efficient number of acres without the threat of a liability penalty.

The economic performance of strict liability and negligence rules is limited in all of the ways in which ambient-based incentives are limited. In addition, producers have uncertainty about whether or not they will be successfully sued for damages, further limiting the effectiveness of this approach. Several factors may influence this uncertainty. For example, the characteristics of agricultural pollution, including dispersion of harm and the inability to identify sources, could make the probability of a producer being sued and held liable very small under strict liability rules. A negligence rule may be more appropriate in these cases because it is not necessary to prove a producer's contribution to damages. Liability based on compliance with 'accepted management practices' would be seen as the fairest, because producers would not be held liable unless they were not in compliance with acceptable practices.

Finally, the litigation process for liability may be expensive relative to other regulatory methods. This expense may prevent individuals from attempting to claim damages, letting polluters go unregulated (Shavell, 1987). Due to these considerations, liability rules are not likely to be first-best and are probably best suited to the control of pollution related to the use of hazardous materials, or to non-frequent occurrences such as accidental chemical spills (Menell, 1990; Lichtenberg, 1992; Wetzstein and Centner, 1992).

Point–Non-point Trading

Pollution trading is gaining increasing acceptance as a cost-effective approach for achieving environmental quality goals (Tietenberg, 1995a,b). The main appeal of trading is its potential to achieve environmental goals at lower social cost than the 'command and control' instruments that have been the dominant approach to pollution control in the US and other developed countries (e.g. Baumol and Oates, 1988; Hahn, 1989; Tietenberg, 1995a,b; Hanley *et al.*, 1997). In 'text book' form, trading systems work as follows. Each polluter receives (through either an endowment, auction, purchase or some other means) a number of pollution permits that specify allowable emissions for the permit holder. Thus, the permit provides the polluter with limited rights to emit pollutants. Through permits, emitters can increase or decrease their allowances by buying or selling with other permit holders, subject to rules governing trades. Pollution sources having greater marginal pollution control costs will purchase permits from sources having smaller marginal costs. The result is that firms having lower control costs emit less, firms having greater control costs emit more, and the maximum total allowable level of pollution is met at lower cost than if trading was not allowed (Montgomery, 1972; Baumol and Oates, 1988; Hanley *et al.*, 1997).

Air- and water-based tradeable permit systems have been implemented in the US, Germany, Canada and Chile to control point sources emitting organic effluents, volatile organic compounds, carbon monoxide, sulphur dioxide, particulates and nitrogen oxides (Tietenberg, 1995a,b). Proposed future applications include an international market for reducing emissions of greenhouse gases (e.g. carbon trading under the Kyoto Protocol). There is also a growing interest in broadening pollution trading schemes to include non-point sources of water pollution (Elmore *et al.*, 1985; Shortle, 1987; Camacho, 1991; Letson *et al.*, 1993; Malik *et al.*, 1993; Rendleman *et al.*, 1995; Faeth, 2000).

Point–non-point trading represents an innovative watershed-based approach to reducing non-point pollution while also improving the cost-effectiveness of the allocation of pollution load reductions between point and non-point sources (Elmore *et al.*, 1985; Shortle, 1987; Camacho, 1991; Malik *et al.*, 1993, 1994; Letson, 1992; Rendleman *et al.*, 1995; Anderson *et al.*, 1997; USDA and USEPA, 1998; Faeth, 2000; GLTN, 2000). Pilot point–non-point trading programmes have been established for the Tar-Pamlico estuary in North Carolina, the Dillon Creek Reservoir in Colorado and Cherry Creek, Colorado. A number of other planned or pilot programmes are being developed (see Table 2.3), often in conjunction with the establishment of total maximum daily loads (TMDLs). Most of these programmes have identified agriculture as the primary non-point source for trading (Table 2.3).

There has been less activity in established US trading programmes than was originally anticipated. A recent assessment indicates that design flaws, rather than problems in the basic concept, are at fault (Hoag and Hughes-Popp, 1997). Given the increasing interest in point–non-point trading, it is important to gain a better understanding of what the programme design options are and how choices among these options can be made to promote cost-effective trading.

While there may be significant potential gains from reallocating pollution control between point and non-point sources, there are also significant challenges in the design of point–non-point trading systems that can realize these gains. Trading between point and agricultural sources entails a fundamental departure from text book tradeable discharge markets (Shortle, 1987; Malik *et al.*, 1993). Because non-point emissions cannot be monitored accurately at reasonable cost and are stochastic, a fundamental issue in the design of agricultural trading programmes is what farmers will trade. Point–non-point systems that have been developed to date involve point sources trading increases in emissions for reductions in estimated loadings from non-point sources (due to the stochastic nature of non-point pollution).[21] Existing and planned programmes work as follows. Point sources are provided with pollution permits, such as through the NPDES system, that define allowable emissions or loadings for the permit holder. These sources would have the option of satisfying the permit on their own, or by purchasing additional allowances or credits from non-point sources. Thus, trading will transfer some control responsibility to non-point sources.[22] However, under existing and planned programmes, agricultural (and other) non-point sources enter into such a commitment voluntarily and are compensated for their abatement efforts.

An alternative to trading mean loadings would be to trade inputs that are correlated with pollution flows (e.g. trading point source emissions permits for agricultural permits restricting the use of polluting inputs such as fertilizers). Systems have also been proposed in which point source emissions could be

Table 2.3. Existing, pilot and planned point–non-point trading programmes in the United States.

Programme	Sources involved	Pollutants traded	Primary non-point sources	PS/NPS trading ratio
Cherry Creek, Colorado	PS/PS and PS/NPS	Phosphorus	Land use projects managed by Cherry Creek Basin Water Quality Authority	Range from 1.3:1 to 3:1
Chesapeake Bay Program (multi-state)	PS/NPS	Nutrients	Agriculture and urban	Greater than 1:1 is suggested to deal with uncertainty
Dillon Creek, Colorado	PS/NPS and NPS/NPS	Phosphorus	Urban, septic, ski areas	2:1
Fox-Wolf Basin 2000 Project, Wisconsin	PS/NPS	Nutrients	Agriculture	Not available
Long Island Sound (multi-state)	PS/PS and (eventually) PS/NPS	Nitrogen	Not yet identified (small % of total loads)	Not available
Lower Boise River, Idaho	PS/PS and PS/NPS	Phosphorus	Agriculture	Site-specific with uncertainty discount built in
Michigan (statewide)	PS/NPS	Nutrients and other	Agriculture	2:1 with site-specific factors
Red Cedar River, Wisconsin	PS/NPS	Phosphorus	Agriculture	Not yet available
Rock River, Wisconsin	PS/NPS	Phosphorus	Agriculture	Site-specific with a base ratio of 1.75:1
Tar-Pamlico, North Carolina	PS/NPS	Nutrients	Agriculture	3:1 for cropland, 2:1 for livestock

Note: Preliminary analyses under way in Ohio, Texas, Maryland, Indiana, Illinois and Virginia.
This list is not exhaustive. Also, some changes are likely given the preliminary nature of some programmes.
Sources: Horan (forthcoming).

traded for reductions in the use of fertilizers and/or reductions of cropland in fertilizer-intensive uses (Hanley *et al.*, 1997).

In addition to the question of what to trade, another fundamental issue in the design of any trading scheme is the rate at which non-point allowances are traded for point source allowances (Shortle, 1987; Letson, 1992; Malik *et al.*, 1993). Because non-point inputs and estimated loadings are imperfect substitutes for point source emissions, trades should not occur at a ratio of one for one. Existing literature provides little guidance, but suggests that factors such as risk and relative contributions to ambient pollution are important in the design of first-best markets (Shortle, 1987; Malik *et al.*, 1993).

Two types of trading system are outlined below. One involves trades of point source emissions for estimated non-point source loadings. The second involves trades of point source emissions for non-point source inputs. Theoretical research has demonstrated that emissions-for-inputs (E-I) trading systems can be designed to provide greater economic efficiency (transactions costs aside) than emissions-for-estimated loadings (E-EL) trading schemes, because they are better able to manage the variability of non-point loads (Shortle and Abler, 1997). The reason, as discussed previously, is that estimated loadings are suboptimal as a basis for non-point pollution control. However, under 'real world' conditions, an E-EL trading system may well outperform an E-I system. We shall return to this issue later.

Emissions for estimated emissions trading

An emissions-for-estimated loadings trading system would consist of two categories of permits: point source permits, $\hat{e}$, and non-point source permits, $\hat{r}$. The former are denominated in terms of emissions while the latter are denominated in terms of estimated loadings. Firms must have a combination of both types at least equal to their emissions, in the case of point sources, or estimated loadings in the case of non-point sources. In existing programmes that include agricultural sources, agricultural sources are not required to have permits. Instead, these sources have an implicit, initial right to pollute, which is consistent with having permits equal to unregulated estimated loadings levels. Trading occurs as non-point sources contract with point sources to reduce estimated loadings in exchange for a fee. Such contracts represent the only enforceable regulations on agricultural sources. However, point sources are ultimately held responsible for meeting water quality goals if they are not met through non-point source reductions (Malik *et al.*, 1994).

In most existing programmes, permits are traded at a rate of 1:1 within source categories and a trading ratio, $t = d\hat{r}/d\hat{e}$, defines how many non-point permits substitute for one emissions permit for trades between source categories. This restriction of 1:1 trading within categories reduces cost-effectiveness when firms' emissions (or loadings) have differential marginal environmental impacts, because uniform trading ratios do not give firms incentives to exploit differences in their relative marginal environmental impacts, as a differentiated system would (Tietenberg, 1995a,b). However, this restriction could provide a net economic gain if it reduces programme administrative and other transactions costs. The same is true for a uniform trading ratio that does not vary depending on the locations of sources involved in a trade. Most existing point–non-point programmes do operate with a single trading ratio, although the ratio is spatially differentiated for a few newer programmes such as the ones in Michigan and Idaho (GLTN, 2000).

Trading ratios for a cost-effective programme, given the 1:1 trading restriction within source categories, have been derived by Horan *et al.* (2000b). As with other studies (e.g. Shortle, 1987; Malik *et al.*, 1993), they found that this ratio can be greater than, less than, or equal to one. A ratio equal to one implies indifference at the margin between the source of control. Ratios in excess of one imply a high cost of non-point control relative to point source control and thus a marginal preference for point source reductions. The reverse is true for ratios less than one.

Little can be said *a priori* about the magnitude of an optimally set trading ratio, though theory suggests that factors such as the relative marginal contributions of point and non-point sources, the degree of environmental risk impacts, correlations between key environmental and cost relationships, and the overall level of heterogeneity associated with point and non-point source could all play a role (Horan *et al.*, 2000b).

Emissions-for-inputs trading

Now consider an emissions-for-inputs (E-I) trading system. As above, we assume two main categories of permits: point (PS) and non-point (NPS). PS permits are denominated in terms of emissions as in the E-EL system. In contrast, NPS permits are differentiated further and denominated in terms of specific inputs. As with the E-EL system, we assume an efficiency-reducing restriction of 1:1 trading of permits within source categories, with trading ratios applicable for trades between source categories and for different inputs. Additional inefficiencies may arise for E-I trading systems where only a subset of inputs are traded, though this is likely to be a practical considera- tion because it will likely be difficult and costly to monitor all inputs (Shortle *et al.*, 1998).

Trades involving pollution-reducing inputs (i.e. those inputs for which increased use reduces pollution) are characterized by some interesting features. Specifically, permits for these inputs may define *minimum* required input use in some situations. When this occurs, then cost-effective trading ratios involving these inputs will be negative (i.e. a reduction in emissions is traded for an *increase* in pollution-reducing inputs) and the economic effect will be to create an opportunity cost associated with reduced use of pollution- reducing inputs.

Cost-effective E-I trading ratios, given the 1:1 trading restriction and the restrictions on the number of inputs requiring permits, have been derived by Horan *et al.* (2000b), though little can be said *a priori* about their magnitudes. As with the E-EL ratio, the E-I ratios can be greater than, less than, or equal to one, and will likely be influenced by similar factors. One difference between E-I and E-EL trading, however, is the impact of input substitution. Specifically, if permits requiring an increase in pollution-reducing inputs also have the effect of increasing the producers' demand for pollution-increasing inputs, then damages could increase as a result. In such cases, trading ratios involving pollution-reducing inputs would be increased to encourage greater control of point sources and to encourage reduced use of pollution-increasing inputs. Accordingly, trading ratios involving pollution-reducing inputs will not necessarily be negative.

Some empirical results

Horan *et al.* (2000a) compared an E-EL system and two E-I systems for point–non-point trading in the Susquehanna River Basin in Pennsylvania, where trades are between farmers and municipal and industrial point sources of pollution. Uniform trading ratios are applied for each system. They found that trading programmes for which non-point permits are defined in terms of loadings are less costly than those based on input use and perform almost as well as the first-best approach. This result occurs largely because loadings are a better indicator of environmental pressures than are inputs. In contrast, programmes in which allowances are defined in terms of non-point inputs are more costly due to the restriction of uniform trading ratios within source categories. This result indicates that differential treatment among sources is likely to improve performance for input-based trading systems, but not for trading systems based on estimated loadings. Of course, the transactions costs associated with increased differentiation of trading ratios are also important to consider, especially given the large numbers of sources often associated with non-point pollution. However, such transactions costs might be justified if the transactions costs associated with the alternative, an estimated loadings-based system, are large relative to those of input-based programmes.

A second result is that the choice of input permit bases can greatly affect the relative performance of input-based trading schemes. In particular, a trading system in which non-point permits are based solely on land use (i.e. placing land in or out of production) does little, if anything, to reduce the expected social costs of pollution. This is because land use has a less direct impact on pollution relative to other inputs, and the effects of changes in land use on pollution depend largely on economic substitution and output effects. This result raises important questions about the heavy reliance on current approaches that focus considerable attention on point sources and extensive margin decisions of non-point polluters (e.g. the CRP).

Thirdly, the majority of control costs fall on non-point sources, indicating that having substantial point source controls relative to non-point controls yields excessive costs. Consider the emissions-for-land trading scheme: significant non-point controls are too costly to undertake in this system but, even so, little is optimally reallocated back to point sources. Instead, the optimal level of control is small and expected social costs are not reduced significantly.

Finally, the trading ratios for E-EL systems are much smaller than those found in existing markets, trading ratios for emissions-for-nitrogen systems are much larger than ratios currently applied in existing markets, and optimal ratios for each scheme can vary considerably depending on watershed characteristics. This result suggests that there may be limitations to using existing markets for guidance for appropriate ratios. Moreover, it suggests that the manner in which environmental performance measures are defined is important. Specifically, as permits are defined for environmental performance measures that are closer to the field and farther from the location of damages (e.g. estimates of field losses), the magnitude of the trading ratio increases optimally.

Commodity Market Distortions and Instrument Choice and Design

Before concluding this chapter, it should be noted that our analysis, and most research on the design of environmental policy instruments for agriculture, generally proceeds under the assumption that agricultural commodity and input markets are not distorted. However, this is not typically the case. Agricultural markets are often distorted by interventions intended to serve farm income, trade, food price, revenue or other policy goals. The distortions take a variety of forms, including output price floors and subsidies, production quotas, input subsidies or administered prices that differ from opportunity costs (Gardner, 1987). These distortions can affect the location, type and severity of agricultural externalities (Hrubovcak et al., 1989; Antle and Just, 1991; Abler and Shortle, 1992; Liapis, 1994; Swinton and Clark, 1994; Plantinga, 1996). An important example is irrigation water pricing in the western United States. In this case, agricultural producers have been charged prices well below the opportunity costs of water, encouraging irrigation practices that are harmful to soil and water quality (Weinberg et al., 1993a; Weinberg and Kling, 1996). Furthermore, the distortions imply that market prices do not measure the social opportunity costs of resources, which has implications for the design of environmental policies. Specifically, when markets are distorted, first-best rules are no longer optimal (Lipsey and Lancaster, 1956). Instead, second-best rules that take into account the effects of policy designs on the costs of market distortions are indicated.

There is a growing literature on the design of second-best environmental policies in the context of imperfect markets. Much of this literature is concerned with the implications of imperfect markets and interactions between environmental taxes with economic distortions introduced by labour and other taxes (Sandmo, 1975; Baumol and Oates, 1988; Oates, 1995; Goulder, 1997). There is also increasing attention on the implications of traditional agricultural policies for the design of agricultural environmental policies. In general, distortions can be expected to affect both the optimal choice of base and the intensity of the application of agricultural environmental policies. For instance, when agricultural commodity prices are supported, consumer prices will exceed marginal costs, implying excess production and associated deadweight costs. Environmental instruments that diminish output will pay an additional dividend by reducing the deadweight costs of the output subsidy (Lichtenberg and Zilberman, 1986). This gain may argue for pursuing second-best agricultural environmental instruments with greater intensity (e.g. higher input tax rates or stricter standards) than would be first-best (Shortle and Dunn, 1991; Laughland, 1994).

Abrahams and Shortle (2000) examined the choices between nitrogen taxes and standards and excess nitrogen taxes and standards for reducing agriculture's contribution to nitrogen pollution under conditions of asymmetric information about control costs and under alternative agricultural commodity price scenarios. They found that economically efficient nitrate policy choices are sensitive to commodity programmes and uncertainty. With the commodity programmes, the most efficient instrument is a fertilizer tax. Without the commodity programmes, the preferred policy is an excess nitrogen tax. Accordingly, the optimal base is affected by the commodity policies.

The implications of commodity policies for the choice and design of environmental instruments and the economic gains from their application will also be affected by the cooperation, or competition, and resulting coordination of commodity and environmental policy choices (OECD, 1989, 1993a; Just and Antle, 1991; Shortle and Laughland, 1994; Weinberg and Kling, 1996; Shortle and Abler, 1999). One approach to coordination, discussed in Chapter 3, is the use of 'green payments' to pursue both environmental and farm income objectives.

Concluding Remarks and Policy Recommendations

Since the early 1980s, economists have been exploring policy instruments for agricultural and other non-point pollution problems. One result has been a body of theoretical research on non-point policy instruments and there are continued developments in this area. This research has identified a catalogue of instruments with theoretically appealing economic properties, but no 'magic bullet'. The sorts of instruments that can in theory bring about a first-best solution to non-point problems are generally too complex, information intensive or costly to implement in practice. There is growing interest in theoretical and empirical research on instruments that make sense in real world conditions.

Empirical research on agricultural non-point pollution control has produced a number of important lessons that can help to guide choices. This literature generally supports the presumption that economic instruments can, in most instances, achieve reductions in pollution from agriculture at a lower cost than regulatory approaches. It also demonstrates that choices about targeting polluters, choices among compliance measures, and the details of the incentives, or regulations, will generally have a significant impact on the economic performance of environmental instruments for agriculture. Other lessons include the importance of watershed-based approaches that coordinate point and non-point controls in watersheds in which both are significant sources, and the importance of coordinating agricultural environmental policies with other agricultural policies.

While much has been learned, significant research issues remain. In particular, more empirical research is needed using integrated modelling approaches at watershed scales that capture multiple sources (e.g. agricultural, urban), multiple stressors (e.g. nutrients, toxic chemicals, suspended solids, pathogens) and multiple environmental endpoints (see Chapter 4). Further, it is important to capture such features as the variability of non-point pollution events. It is also important to incorporate transactions costs. Most analyses of second-best instruments acknowledge transactions costs to be important, yet few studies actually model these costs explicitly. Those that do model transactions costs do not do so in a consistent fashion, as these costs are assumed to depend on one set of factors in one study and other factors in other studies. This is not too surprising, given that the empirical literature on estimating transactions costs is extremely limited and it is not yet clear on what these costs will depend. Additional research is needed to assess the normative implications of transactions costs (Krutilla, 1999) as well as to quantify them (e.g. Carpentier et al., 1998; McCann and Easter, 1999).

Endnotes

1. This chapter borrows and builds on our previous joint work and work with others. In particular, see Shortle and Abler (1997), Shortle et al. (1998), Weersink et al. (1998), Ribaudo et al. (1999) and Shortle and Griffin (2001)).
2. Nature can be a large source of some pollutants that are also contributed by anthropogenic agricultural sources. Uncertainty about agricultural contributions to acidification, nutrient enrichment and other problems is due in part to limited information about the contributions from natural sources (Chesters and Schierow, 1985).
3. See Abler and Shortle (1991a), Phipps (1991), Braden and Segerson (1993), Dosi and Morretto (1993), Shogren (1993), Malik et al. (1994) and Tomasi et al. (1994) for useful discussions of information issues in non-point pollution control and their policy implications.
4. One approach to dealing with non-point pollution in some instances is to convert diffuse non-point pollution into point source pollution. This is clearly limited to

cases where pollutants can be collected for discharge at a central point at reasonable cost.

5. Indeed, many of the emissions taxes that are implemented in practice for point sources are in fact taxes on emissions proxies (OECD, 1994a).

6. Ours is a slightly more complex model than the one adopted by Griffin and Bromley, for consistency in our discussion of more recent models below.

7. More accurately, an efficient solution would also take entry and exit into consideration through the choice of n. Griffin and Bromley and Shortle and Dunn did not consider this important issue. We delay our discussion of entry/exit issues until later.

8. The environmental objective in Griffin and Bromley's model was to restrict a simple aggregation of non-point emissions, i.e. using our present notation, $W = \Sigma r_i + \Sigma e_i$.

9. We have made one simplification relative to Shortle and Dunn's model in that we do not model asymmetric information between farmers and the regulatory agency about farmers' profit functions. We return to this issue in a later section.

10. A special case of cost-effective outcomes is that of *ex ante* efficiency, which occurs when environmental performance is measured by expected damages (i.e. $W = D$) and the performance target is the level of expected damages that arises in the efficient solution (i.e. $T = D^*$).

11. The relation for expected after-tax profits is identical to the expected after-tax profits for a producer who faces a runoff tax. Accordingly, risk-neutral producers will respond the same to instruments based on runoff as to those based on estimated runoff because, in each case, decisions are made *ex ante* (e.g. before weather and its impacts are known), based on the expectation of runoff. This discussion therefore also applies to runoff-based instruments, which could be utilized in the future if technologies are developed to monitor runoff cost-effectively.

12. Let $f = f(q)$ (f', $f'' > 0$), where $q = q(h, v)$, h is deterministic and v is a stochastic variable. Then $cov\{f'(q), \partial q/\partial h\}$ is of the same sign as $cov\{q, \partial q/\partial h\} = 0.5(\partial var\{q\}/\partial h)$, where this equality follows from: $\partial var\{q\}/\partial h = \partial(E\{q^2\} - E\{q\}^2)/\partial h = 2(E\{q\partial q/\partial h\} - E\{q\}E\{\partial q/\partial h\}) = 2cov\{q, \partial q/\partial h\}$. If $f'' < 0$, then $cov\{f'(q), \partial q/\partial h\}$ will have the opposite sign relative to $\partial var\{q\}/\partial h$. This result is used throughout the paper, although with different definitions for f, q and h.

13. The site with the smallest expected incremental net private social benefit relative to expected incremental external costs is defined as the marginal site. Other sites that optimally remain in production are defined as infra-marginal, and sites on which production optimally does not occur are defined as extra-marginal.

14. Alternatively, κ_i could be a subsidy applied to extra-marginal farms that do not enter into production. Because the subsidy is larger than their after-tax profits, extra-marginal farms will find it more profitable not to produce. This type of subsidy would be similar to US Conservation Reserve Program subsidies paid to farmers to take land out of production.

15. At the farm level, a tax on the use of one input (e.g. a chemical) may increase the demand for alternative, non-targeted inputs (e.g. other chemicals) that could also be harmful to the environment. At the market level, a tax could impact input and output prices and alter the demand for non-targeted inputs in ways that could be environmentally damaging. It is optimal to consider such impacts when choosing among bases and setting instrument levels so that any adverse impacts are not

too great. For example, a first-best tax would be negative for inputs that reduce emissions. However, the optimal second-best tax for a pollution-decreasing input will be positive if an increase in the use of the input is associated with increased demand for the use of pollution-increasing inputs, resulting in adverse environmental consequences. For further discussion of substitution effects, see for example Hrubovcak *et al.* (1989), Bouzaher *et al.* (1990), Eiswerth (1993), Braden and Segerson (1993), Schnitkey and Miranda (1993), Shogren (1993), Claassen and Horan (2001).

16. The value of information in this context is the expected increase in the expected net benefits that would result from improved policy design.

17. A related literature investigates the design and performance of group-based instruments or contracts (under conditions of asymmetric information) in which economic penalties or rewards are based on the performance of a group of polluters. For example, see Xepapadeas (1991, 1994), Herriges *et al.* (1994), Govindasamy *et al.* (1994) and Byström and Bromley (1998).

18. Implementation of state-dependent ambient taxes is not likely to be significantly more demanding than that of state-independent tax schemes. To aid producers in their decisions, each firm could be provided with a schedule of tax rates (for the linear case) or tax bills (for the non-linear case) corresponding to different realizations of the random variables. This is not greatly different from that of graduated income taxes. Income taxpayers know the rules that will be used to determine their taxes (or at least they are legally presumed to know them), but the actual base and rate are uncertain.

19. It is possible to develop a second-best uniform, linear ambient tax that is state-independent. The optimal tax rate would equal expected marginal damages, plus the average covariance between marginal damages and marginal ambient levels from input use, normalized by average expected marginal ambient pollution levels. However, it is not immediately apparent that transactions costs would differ from the efficient, state-dependent taxes. This is because the state-dependent taxes share the same tax base and utilize the same information for (optimal) rate design as other ambient taxes. How the taxes would perform in a second-best world is an interesting question, but inherently empirical.

20. Shavell (1987) discussed circumstances for which publicly and privately initiated approaches are most appropriate.

21. The liability rule may also be influenced by the relationship between polluters and the victims, defined as either unilateral care or bilateral care (Segerson, 1995). Unilateral care is a situation in which only the polluter influences damages; in other words, the victim has no way of protecting himself. Alternatively, it is sometimes possible for the victim to protect himself. For example, the victim may be able to purchase a filtration system to protect against contaminated groundwater. This situation is known as bilateral care. Under some rules, liability is not assessed to polluters if the victim failed to take reasonable preventative actions (Segerson, 1995).

22. We use the term loadings as opposed to runoff because this term is commonly used in reference to point–non-point trading programmes. For our purposes, loadings can essentially be thought of as runoff. In practice, however, the two concepts are different as loadings are often used to define how much of a pollutant enters or loads into river reaches or the like, while runoff often refers to edge-of-field losses.

23. Although responsibility is transferred, liability often is not. In most programmes such as the Tar-Pamlico programme or the programme proposed for Michigan, point sources are ultimately held liable for whether or not the conditions of the permit are satisfied. The point source is liable for any remaining reductions if a non-point source does not take appropriate abatement actions, although the point source may then have the right to sue the non-point source for compensation. Even so, failure to transfer liability to non-point sources when a trade occurs may represent an important barrier to trade.

Chapter 3

Voluntary and Indirect Approaches for Reducing Externalities and Satisfying Multiple Objectives[1]

RICHARD D. HORAN[1], MARC RIBAUDO[2] AND DAVID G. ABLER[3]

[1] *Department of Agricultural Economics, Michigan State University, East Lansing, MI 48824-1039, USA;* [2] *USDA Economic Research Service, Washington, DC 20036-5831, USA;* [3] *Department of Agricultural Economics and Rural Sociology, Pennsylvania State University, University Park, PA 16802, USA*

Chapter 2 focused on enforceable instruments for inducing farmers to adjust their production and pollution control practices, thereby impacting water quality directly. While enforceable instruments have found some use as a means for reducing water pollution from agriculture, governments have relied largely on voluntary compliance approaches that combine public persuasion with technical assistance to encourage and facilitate adoption of environmentally friendly technologies (OECD, 1993a, 1998). This is due in part to the difficulties involved with designing and administering environmental policies for agriculture, difficulties that are described elsewhere in this book. It is also due to the political clout of agricultural producers in most developed countries. This political clout manifests itself in agricultural price and income support programmes that the OECD (2000) estimates cost US$361 billion per year in the Organisation for Economic Cooperation and Development (OECD) member countries, or 1.4% of OECD gross domestic product (GDP). Cutbacks in these policies could lead to less intensive production practices and could help to cut agricultural pollution (Abler and Shortle, 1992; OECD, 1998), but governments worldwide have been reluctant to undertake agricultural policy reform.

This chapter examines four types of voluntary policy instrument that have indirect impacts on water quality.[2] These are education, research and development (R&D), green payments and conservation compliance. The impacts of these instruments are indirect in that the instruments neither require (e.g. via regulations) nor directly induce (e.g. via an unavoidable tax on polluting activities) changes in behaviour. Instead, the instruments initiate a chain of actions that, in the end, are intended to induce farmers to change production practices voluntarily and thus improve water quality. While these policy approaches are indirect, they have been at the forefront of government efforts to reduce water pollution from agriculture.

Why should governments use indirect approaches when the direct approaches presented in Chapter 2 are available? One reason, alluded to above, is political constraints on instrument choices. There are also economic reasons related to the supply of research and development on new environmentally friendly technologies, and the economics of farm-level information acquisition. Finally, most nations have a variety of policy interests related to agriculture. These include farm income, trade, food security, and food price and safety objectives. Mitigating the adverse water quality impacts of agriculture is a comparatively new concern. Policy developments in this area take place within the broader agricultural policy context. Historical concerns and approaches to farm problems shape choices of policies for addressing emerging environmental issues.

Education

Agricultural non-point pollution problems often involve small producers who, because of their size and the fixed costs of acquiring information, may not invest much in information on techniques for limiting water pollution. Public agencies may have significantly better information than producers about pollution control or pollution prevention practices. Disseminating such knowledge could provide environmental improvements if this knowledge encouraged producers to operate in more environmentally friendly ways – either with existing methods and technologies or by adopting alternative technologies.

Education programmes are an important part of non-point pollution programmes in many countries and will likely remain so. In the US, for example, education plays a significant role in every state and federal non-point source water quality programme, most recently in the Clinton Administration's Clean Water Action Plan (USEPA, 1998c; Nowak *et al.*, 1997). Education programmes supply producers or consumers with information on practices for reducing non-point pollution, technical assistance to facilitate adoption, and encouragement to adopt, out of either self-interest or concern for broader societal well-being. Common mechanisms for conveying information to farmers include demonstration projects, technical assistance, newsletters, seminars and field days.

Education is popular for a number of reasons. It is less costly to implement than many other programmes, and the infrastructure for carrying out such a programme is largely in place (e.g. through agricultural extension offices). In addition, there is some empirical evidence that education can be effective in getting farmers to adopt certain environmentally friendly practices (Gould *et al.*, 1989; Bosch *et al.*, 1995; Knox *et al.*, 1995). Education programmes are most effective when encouraging 'environmentally friendly' actions that also happen to be profitable (Feather and Cooper, 1995; Musser *et al.*, 1995). Indeed, educational assistance is often seen as a means of achieving so-called win–win solutions to water quality problems, whereby net returns and water quality are both improved (USEPA, 1998c).

Some practices that could help to protect and enhance water quality and that have been shown to be more profitable than conventional practices in many settings include conservation tillage, nutrient management, irrigation water management and integrated pest management (Fox *et al.*, 1991; Conant *et al.*, 1993; Bull and Sandretto, 1995; Ervin, 1995). Extension education to encourage and support the adoption of integrated pest management (IPM) is a key component of pesticide control programmes in Canada, Denmark, The Netherlands, the USA and Sweden. Basic IPM practices, such as pest scouting to detect whether it would be profitable to apply pesticides, are now used by a significant proportion of US farmers, though use of more sophisticated IPM techniques has been limited (Fernandez-Cornejo *et al.*, 1998). Pesticide use in Sweden has been greatly reduced since the early to mid-1980s through a mixed approach of regulations, incentives and education, along with fortuitous developments in low-dose application and spraying pesticide technology (Weinberg, 1990; Bellinder *et al.*, 1993; Pettersson, 1994). The independent contribution of education is unclear. Agricultural extension programmes in most OECD countries now include environmental components (OECD, 1989, 1993a).

Education when producers only care about profitability

Opportunities for simultaneously increasing water quality and farmer profits would seemingly make water quality protection easy to accomplish. However, even though education may encourage a producer to operate along a more socially efficient production frontier, private and public objectives will still generally differ. As long as producers only consider profitability, competition will generally drive them to operate in ways that will not necessarily coincide with environmental goals. In fact, it is entirely possible that providing education about production practices might even lead to changes in the scale of production and input mix that cause water quality to worsen (Ribaudo and Horan, 1999).

These arguments can be illustrated graphically using a simple example involving a single farm (which may be one of many contributors to non-point pollution in a watershed) in which a single input leads to water quality impairment, and pollution creates no on-farm costs. The relationships between production and expected water quality are depicted graphically in Fig. 3.1. We say *expected* water quality because actual water quality is stochastic due to the weather, particularly precipitation. Quadrant I illustrates

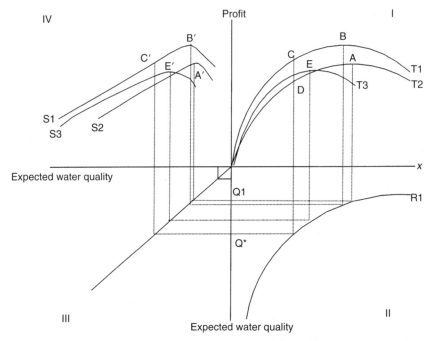

Fig. 3.1. Profit maximizing production decisions and their effects on water quality.

the relation between input use and the producer's net returns (i.e. the restricted profit function). Without loss of generality, the Profit axis could be thought of as the expected utility of profits for risk-averse producers when there is production uncertainty. Quadrant II illustrates the relationship between input use on the farm and expected water quality, taking the actions of all other non-point polluters as given. Quadrant III projects the relations from Quadrant II into Quadrant IV, which depicts the relation between expected water quality and net returns. The way in which producers account for water quality in their production decisions is reflected in this quadrant.

Define Q^* to be the Pareto or economically efficient level of expected water quality (with production occurring at point C on curve T1). Expected water quality levels will be below Q^* when: (i) producers do not consider the economic impacts of their production decisions on water quality, and/or (ii) producers face uncertainty or have a limited understanding of the production and environmental impacts of their management choices. The purpose of educational programmes is to reduce producers' uncertainty and to improve their knowledge about production and environmental relationships (both for production technologies they are currently using and for alternative technologies). Proponents of such programmes believe expected water quality will be improved if the information provided encourages producers to: (i) consider the environmental impacts of their choices, and/or (ii) simultaneously improve expected water quality and profitability by using existing technologies more efficiently or by adopting alternative, more environmentally friendly technologies (Nowak *et al.*, 1997). Assuming that the producer's only objective is to maximize profits (and thus has no altruistic motivations), only (ii) has any chance of success.

For simplicity, assume that the only source of uncertainty relates to the production frontier, which may lead producers to use inputs inefficiently.[3] This situation is represented by curves T1 and T2 in Fig. 3.1. T2 reflects the production technology that the producer is currently using (i.e. the set of practices in use) and the skill with which it is being used. T1 reflects the government agency's expectations about the production frontier, which are assumed to be more accurate than the producer's due to better information from publicly supported research.[4]

Initially consider the case in which the producer's only objective is to maximize profits. A profit-maximizing producer will produce (inefficiently, according to T2) at point A in Fig. 3.1. In the absence of any outside programmes or intervention, the producer would not voluntarily move to point D so that economic efficiency is achieved, since net returns would be reduced without any compensating private benefits. In Quadrant IV, profit maximization indicates that the producer operates at point A′. It would be pointless to educate the producer about the relationships between production and water quality because the producer has no altruistic or stewardship motives. However, by educating the producer about the frontier T1, where profits are higher for each level of input use, the producer could be encouraged to use existing management practices more efficiently or to adopt alternative practices so that he/she operates along T1.[5] Once on T1, the producer could operate at the efficient point C to meet the expected water quality goal and at the same time increase net returns relative to operation at point A on T2 (though there may be values of C for which net returns might be reduced). Such an outcome appears to be a win–win solution for the farmer. However, even though producing along a more socially efficient production frontier, the producer's goals will still generally differ from society's goals. As long as producers only consider profitability, competition will drive the producer to operate at point B (note that point C is necessarily to the left of B). The expected water quality levels that correspond to B are an improvement over the initial situation when the producer produced at A, but are still less than efficient levels. Thus, educational assistance alone is not enough to ensure that the water quality goal is met.

It is entirely possible that providing education about production practices might even reduce expected water quality. Suppose the producer originally produced according to T3, so that profits were maximized at E. After receiving educational assistance, the producer would have an incentive to produce at point B on T1. Net returns increase in this case, but so does the use of input x. The result is that expected water quality is worse than it was before education was provided. This result is more than just a curiosity. For example, there is evidence that some IPM practices have actually increased the amounts of pesticides farmers use (Fernandez-Cornejo *et al.*, 1998).

Education when producers care about the environment

Research has demonstrated that farmers are well informed of many environmental problems, and that most US farmers perceive themselves to be stewards of the land (Camboni and Napier, 1994). Educational programmes could take advantage of altruism or land stewardship motives by informing farmers about local and on-farm environmental conditions and about how a change in management practices could improve local and on-farm water quality.

Figure 3.2 considers the case where farmers have some altruistic or land stewardship motives that influence their decision-making. It is possible in this case to construct a utility indifference map showing the rates at which a producer is willing to trade net returns for increased water quality. The point along the water quality/net returns frontier (Quadrant IV) where a producer will operate is at the point of tangency with an indifference curve (e.g. U1 or U2), or where the marginal rate of substitution (MRS) between net returns and water quality is equal to the slope of the net returns/water quality frontier. At this point, the producer's utility is maximized.

Suppose an altruistic producer does not believe he or she is contributing to water quality problems and is not aware of T1. Production will initially take place along T2 at A (or at A' in Quadrant IV). Since the producer is unaware of R1, the MRS between net returns and water quality is 0. Suppose

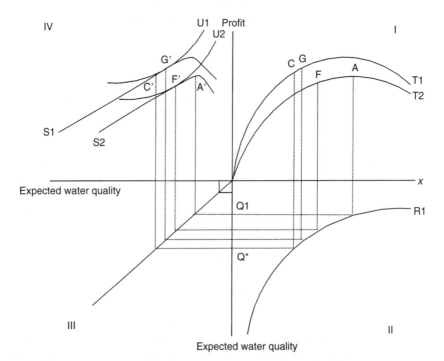

Fig. 3.2. Utility maximizing production decisions.

that the producer is informed of how the use of x is affecting water quality so that the relationships expressed by R1 and S2 are revealed to the producer. The response of the producer to this information will depend on their willingness to give up some net returns to protect water quality, expressed by the indifference curves in Quadrant IV. Production on T2 will now occur to the left of A, at F (F' in Quadrant IV), where indifference curve U2 is tangent to S2. In the example, water quality is improved and utility increased (point A' lies below U2). This is a win–win situation for the producer, even though net returns are reduced.

Suppose now the producer is educated about T1. The altruistic producer will have an incentive to make production decisions based on the trade-offs defined by S1 and U1, operating now at point G. In this example, both water quality and net returns are higher than for points A and F, a win–win situation. However, this need not be the case. The ultimate impacts on water quality will generally depend on the nature of T1 and R1 relative to T2, and on the MRS between net returns and water quality. If expected water quality does improve as a result of education, the degree of improvement relative to Q* depends on how strongly the producer values environmental quality. Efficiency is obtained only for the special case in which each producer makes production decisions while fully internalizing their marginal contribution to expected environmental damages.

Farmers have been shown to respond to education programmes when their own water supply is at stake (Napier and Brown, 1993; Anderson *et al.*, 1995; Knox *et al.*, 1995; Moreau and Strasma, 1995). However, the perceived benefits to farmers of significant changes in existing practices are often small (Beach and Carlson, 1993; Norton *et al.*, 1994). Similarly, experience with education programmes and empirical evidence indicates that altruism or concern over the local environment plays only a very small role in farmers' decisions to adopt alternative management practices (Camboni and Napier, 1994; Franco *et al.*, 1994; Weaver, 1996). Agricultural markets are competitive and market pressures make it unlikely that the average farmer will adopt costly or risky pollution control measures for altruistic reasons alone, especially when the primary beneficiaries are downstream (Bohm and Russell, 1985; Nowak, 1987; Abler and Shortle, 1991a; Napier and Camboni, 1993).

A basic requirement for altruism to be the motivating factor for change is that farmers believe there is a problem that needs to be addressed, and that their actions make a difference (Padgitt, 1989; Napier and Brown, 1993). Surveys of producer attitudes and beliefs toward the relationship between their actions and water quality consistently find that farmers generally do not perceive that their activities affect the local environment, even when local water quality problems are known to exist (Hoban and Wimberly, 1992; Lichtenberg and Lessley, 1992; Pease and Bosch, 1994; Nowak *et al.*, 1997). For example, USDA's Water Quality Demonstration Projects did not significantly change farmers' perceptions about their impacts on water quality even though the projects were located in areas with known water quality problems

(Nowak *et al.*, 1997). One could easily imagine that convincing farmers of their contribution to a non-point source pollution problem may be difficult, especially given that non-point source pollution from a farm cannot be observed and that its impacts on water quality are the result of complex processes and are often felt downstream from the source. If there are many other farmers in the watershed, a single farmer may, justifiably, believe that his/her own contribution to total pollution loads is very small. Even if farmers do take appropriate actions to improve water quality, they generally will not be able to observe whether these changes in management actually improve water quality. Farmers will have to take as a matter of faith any information provided about the water quality impacts of changes in their production practices.

Education as part of a more comprehensive policy

There is ample evidence that public perceptions about environmental risks are often at odds with expert assessments and that people do not necessarily respond to risk information in ways that experts consider logical (e.g. Fisher, 1991; Lopes, 1992). To the extent that information programmes are used in an attempt to change producer behaviour, it is important that they be designed with a good understanding of the kinds of message and delivery mechanism that will have an impact on the target audiences.

Education's greatest value may be as a component of a pollution control policy that relies on other tools. By providing the information that farmers need for efficient implementation of changes in production practices, overall pollution control is attained at lower cost.

Research and Development (R&D)

There has been a tremendous growth of interest in recent years in 'alternative', 'green', 'sustainable' and 'environmentally friendly' agricultural technologies. Probably the best-known alternative technology is integrated pest management (IPM), which involves the use of economic thresholds to determine when pest populations are approaching the point where control measures may be profitable. A variety of control measures may be used, including not only pesticides but also releasing sterile insects to mate with fertile ones, spraying insects with synthetic hormones to prevent their development, and releasing 'beneficial' insects (predators, parasites or pathogens). Other alternative technologies include crop rotations optimized for specific locations (through adjustments in crops planted, rotation length, and tillage, cultivation and fertilization practices), improved manure storage and application practices, and more precise fertilizer and pesticide application techniques. Of course, plant breeders regularly release new seed varieties having desirable properties such as improved disease and pest resistance.

Biotechnology is another important alternative technology. It is already having significant impacts on agricultural production in many countries and could lead to revolutionary changes in the types of crops and livestock produced and the ways in which they are produced (Fernandez-Cornejo and McBride, 2000). Plant biotechnology has the potential to yield crops with significantly greater resistance to a whole host of pests and diseases, necessitating fewer insecticides and herbicides. Work is under way to engineer pest vectors into beneficial insects as part of IPM strategies. Perhaps the most promising plant biotechnology from an environmental perspective, though years if not decades away, is nitrogen-fixing cereal varieties. These varieties would fix atmospheric nitrogen in a manner similar to legumes, which could dramatically reduce nitrogen fertilizer usage. Genetically modified organisms (GMOs) with tolerance to specific herbicides have also been developed. Concerns have been raised that these may promote herbicide usage, although that has not happened to date (Heimlich et al., 2000). Animal biotechnology has the potential to yield livestock that process feed more efficiently, leading to reduced feeding requirements and fewer nutrients in animal wastes. Feed may also be genetically modified so as to reduce nutrients in livestock wastes.

Economic responses to new technologies

Analyses of the environmental impacts of potential new agricultural technologies often focus on their biological, chemical and physical properties relative to existing technologies (e.g. National Research Council, 1989; Logan, 1993; OECD, 1994d). These analyses typically endeavour to assess environmental externalities associated with production of a given tonne of output, or production on a given hectare of land, using new technologies versus existing technologies. For example, how much of a given herbicide is required to produce a kilogram of a new maize variety versus an existing variety? Alternatively, what is the yield of wheat under no-till versus conventional tillage? These kinds of question are critical but, by themselves, they do not tell us the environmental impacts of new technologies, because they do not take into account the economic responses of producers and consumers to new technologies.

One key economic consideration is, of course, adoption. To have an impact, new technologies must be adopted. If they are to be adopted voluntarily, they must be expected to be profitable to producers. If use is mandated by law, then political acceptability and cost-effectiveness considerations would in most situations require any negative impact on producers to be small (Abler and Shortle, 1991a). However, widespread adoption is only one economic consideration.

Environmentally friendly technologies can be broadly classified as either pollution prevention or pollution abatement. Pollution prevention (Freeman, 1993) is

> the use of materials, processes, or practices that reduce or eliminate the creation of pollutants or waste at the source (e.g. no-till). Pollution prevention includes the practices that reduce the use of hazardous materials, energy, water, or other sources and practices that protect natural resources through conservation or more efficient use.

Pollution abatement, by contrast, involves 'end-of-pipe' solutions and other methods of treating pollutants once they have been created (e.g. buffer strips). Of course, there are many types of technical change that do not fit into either of these two categories, and other types may have both pollution prevention and pollution abatement characteristics. Current interest, both in agriculture and in other sectors, is centred heavily on changing production processes so as to prevent pollution in the first place rather than finding better ways to clean it up after the fact.

Pollution prevention technologies could be viewed in at least two ways. First, they could be viewed as new methods of production that completely eliminate the need for one or more polluting inputs. One could think of many innovations that fall under the latter case – for example, 'no-till' farming has been adopted by a significant number of US grain producers, eliminating the use of tillage equipment that had contributed to soil erosion on those farms (USDA ERS, 1997). Provided these technologies are economically attractive enough to be adopted by producers, they will improve environmental quality. Alternatively, one could see pollution prevention technologies as reducing the quantities of one or more polluting inputs required to produce any given level of output without making total elimination of those inputs profitable. This second case is more environmentally ambiguous.

Profit-maximizing producers will not voluntarily adopt new production processes unless they are less expensive than existing processes. If they are less expensive, they will be adopted and marginal cost will fall. At the market level, competition among producers will pass the cost reduction along to consumers in lower prices, which will stimulate the quantity of output demanded. This increase in output demand will work to raise the derived demand for all factors of production, including those associated with pollution. At a minimum this implies a smaller reduction in pollution than would be obtained if output were held constant. If the increase in output were large enough, the total use of polluting inputs could actually rise, even though input usage per unit of output would fall. Simulation analyses by Abler and Shortle (1995, 1996) and Darwin (1992) indicate that such a scenario is not merely possible but plausible.

The results from Abler and Shortle (1995, 1996) suggest that, in general, two conditions must be met in order for total usage of environmentally damaging inputs to go down. Firstly, the alternative technology must be a good substitute for environmentally damaging inputs. Secondly, the demand for the agricultural product in question must not be too price-elastic. The first condition is necessary to reduce significantly the usage of environmentally damaging inputs per unit of output. The second condition is necessary to hold in check the increase in product demand, and in turn output.

Pollution abatement technologies have no impact on production relationships, but they do reduce pollution for any given level of input usage. For example, buffer strips may be made more effective in filtering out nutrients before they reach waterbodies. The environmental benefits of pollution abatement technologies are reasonably obvious. However, because these technologies typically impose costs on users, farmers are unlikely to adopt them unless they are provided with financial incentives or are required to do so.

Environmental policy and incentives for R&D

The section above considers responses by producers and consumers to pollution prevention and pollution abatement technologies once they have been developed. In this section we consider how environmental policies for agriculture might affect incentives in the public and private sectors to conduct R&D on pollution prevention and pollution abatement technologies. In the US, both public and private agricultural research have historically been biased in favour of chemical-intensive techniques (e.g. Antle, 1984; Huffman and Evenson, 1989; Fawson and Shumway, 1992). It is widely acknowledged that similar forces have been at work in the EU. For example, Becker (1990a,b) found that technical change in Germany has been fertilizer- and breeding stock using (Chambers, 1988).

The economic literature on the development and adoption of environmentally friendly technologies dates from the late 1970s. Magat (1978, 1979) was one of the first authors to investigate how environmental policies affect the types and speed of firms' innovations. Downing and White (1986) extended Magat's work to include additional policy instruments and to examine the implications of different strategies on the part of the regulatory authority. Milliman and Prince (1989) and Jung *et al.* (1996) ranked several policy instruments in terms of their potential to induce the development and adoption of pollution abatement technologies. In general, these studies found that emissions taxes or other schemes that put a 'price' on pollution, such as marketable pollution permits, provide the greatest incentives for environmentally friendly R&D. The reason is that firms can reduce their tax liability or expenses on pollution permits by developing technologies that reduce their emissions. On the other hand, schemes that provide the producer with no financial incentives to reduce pollution beyond a certain point – such as emissions standards or technology standards – provide few incentives for environmentally friendly R&D. The producer has incentives to develop technologies that help to reduce the cost of meeting the standard, but there are no incentives to develop technologies that reduce emissions beyond what the standard requires.

One important limitation of these studies is that they consider firms that do all their own R&D. In agriculture, hardly any commercial farms are involved in research. Nearly all farms rely on technologies developed by input suppliers (seed companies, farm equipment companies, etc.) and government agencies. Input suppliers have different objectives than farmers. They seek to maximize profits from the products and services that they provide, not profits from sale of farm commodities. Environmental policies imposed on farmers will affect the amount and direction of R&D by input suppliers only to the extent that there are good market linkages between farms and input suppliers. This condition holds in developed countries but not in many developing countries. Zilberman *et al.* (1997) found that improper incentives associated with socially suboptimal input prices led to socially inefficient investment (and adoption) of precision technologies that would require fewer inputs.

In the case of government agencies, there are no direct market linkages to transmit signals from farmers to researchers. Signals must instead be transmitted through public institutions such as agricultural extension and through political channels. This is problematic because the signals may become 'noisy' and may compete with other signals that researchers are receiving. For example, simulation analyses by Abler (1996) indicated that a tax on fertilizers and pesticides in US maize production could lead to environmentally beneficial changes in private-sector research, but that these benefits could be muted by offsetting changes in public-sector research. The reason was that none of the politically important groups with respect to public-sector maize R&D decisions in the US stood to gain from research that reduced the use of polluting inputs. In related work, simulation analyses by Shortle and Laughland (1994) indicated that a tax on fertilizers and pesticides in US maize production could lead to offsetting adjustments in farm price and income support policies (in order to mollify farm constituencies) that greatly reduced the environmental benefits of the tax.

R&D as part of a more comprehensive policy

R&D cannot stand on its own as a water pollution control tool, because technology is only one component of water quality improvement. Even with the most efficient, environmentally friendly technology, farmers will still have incentives to over-apply inputs that contribute to non-point source pollution, because the off-farm costs of pollution do not show up on the farmer's bottom line. However, R&D can be a valuable component of other approaches that provide farmers with more direct incentives to reduce non-point source pollution. Indeed, one can find examples from many countries of how environmental policies have encouraged the development of new production processes, new products and even entirely new industries (e.g. Caswell *et al.*, 1990; Kemp *et al.*, 1992; Porter and van der Linde, 1995).

Green Payments

Environmental policies are typically based on the 'polluter pays' principle, which holds that polluters themselves should bear the costs of actions taken to protect the environment. Agriculture's political clout in most countries makes it unlikely that this principle will be applied to agriculture, at least not in the near future. Instead, policies based on the 'pay the polluter' principle seem more likely (Hanley *et al.*, 1999).

Agricultural policies have been reformed in recent years in the US, EU and other countries, but the pressure to provide some type of support to the agricultural sector remains strong. The only real questions concern the level of support and how future support payments will be structured. One option is a system of 'green payments' that could improve the environmental performance of agriculture and at the same time provide income support to agricultural producers (Batie, 1994; Heimlich, 1994; Lynch and Smith, 1994; Smith, K.R., 1995; Wu and Babcock, 1995, 1996). For example, the recently proposed Conservation Security Act in the US (S.3223, HR.5511) would provide payments to farmers based on the adoption of designated conservation practices. Green payments appeal to many people because they are voluntary and because they appear to offer an environmentally friendly alternative to traditional agricultural price and income support policies.

Green payments can be defined as payments to producers based on either actions taken to reduce non-point source pollution or on the probable environmental results of such actions. While payment levels may be determined by a number of factors, such as non-point pollution goals or equity considerations, the basis (i.e. management choices or environmental performance measures) to which they are applied effectively determines whether or not a payment is 'green'. For example, a payment to reduce nitrogen use is a green payment – even if the payment is set at levels such that its primary intent is to transfer income. Alternatively, a payment based on something unrelated to emissions would not be a green payment.

Green payments are not a new concept. In the US, state and federal agri-environmental programmes, such as the Conservation Reserve Program (CRP) and Water Quality Incentives Program (WQIP), have relied heavily on green payments, though their success has been limited in a number of dimensions. Firstly, their environmental impacts have been limited because agriculture remains a major contributor to non-point source pollution. Secondly, the cost-effectiveness of these programmes in achieving non-point pollution reductions is suspect, due to a lack of specific environmental goals and a lack of attention to critical design issues related to non-point pollution control. Finally, these programmes have probably had little impact on farm income, with the exception of the CRP. If there is to be a continued or expanded use of green payments to reduce agricultural non-point pollution, additional research is needed on how they can be designed to achieve economic and environmental goals with maximum economic performance.

In general, green payments programmes can be designed in a number of ways, with each differing along several important dimensions. Programmes may have different environmental and income objectives, different degrees of cost-effectiveness, and may imply different economic trade-offs between and among consumers, producers and the environment. Under a traditional welfare economics approach, environmental goals could be achieved cost-effectively using traditional environmental policy instruments (Chapter 2). Equity objectives would then be efficiently addressed using non-distortionary policies (i.e. lump-sum transfers) to adjust the income distribution. However, contemporary welfare economics rejects the feasibility of lump-sum transfers to achieve equity objectives, and suggests considering both efficiency and equity in optimal policy design (e.g. Gardner, 1987).

Given that non-distortionary transfers are impossible, efficiency and equity objectives can be usefully modelled by a social objective function, U (Gardner, 1987). Following Horan et al. (1999a), we define U over the economic welfare of relevant groups: consumers, producers, owners of factors of production, those damaged by non-point pollution, and taxpayers. The associated welfare measures are consumer's surplus (CS), producer quasi-rents $(PS_i, i = 1, ..., n$, where i indexes farms), factor surplus (FS), pollution damages $(D_k, k = 1, ..., K$, where k indexes damage sites) and net government transfers (G). Thus, $U = U(CS, PS_1, ..., PS_n, FS, -D_1, ..., -D_K, -G)$, with U increasing in all its arguments. U is random because stochastic variations in weather, particularly precipitation, cause environmental damages to vary for any given set of management choices by producers. (Other sources of randomness could also be modelled.) How efficiency and equity objectives are satisfied depends on the choice of instruments, how the instruments are designed and implemented, and market and information structures. For instance, green payments are optimally chosen, designed and implemented to maximize expected utility, $E\{U\}$, subject to producer responses.

To illustrate the trade-offs involved with choosing instruments to maximize $E\{U\}$, consider an input subsidy scheme with subsidies of the form $s_{ij}(x_{ij}^0 - x_{ij})$ for all i and j, where x_{ij} is the jth input used on the ith farm, s_{ij} is the corresponding subsidy rate and x_{ij}^0 is the baseline input level from which the subsidy is evaluated. No payments are made when $x_{ij} > x_{ij}^0$ for pollution-increasing inputs or when $x_{ij} < x_{ij}^0$ for pollution-decreasing inputs. We have $s_{ij} > 0$ for pollution-increasing inputs and $s_{ij} < 0$ for pollution-decreasing inputs.

An efficient subsidy rate equals an input's expected marginal contribution to damages from a particular farm (Chapter 2). With equity concerns, however, optimal subsidy rates reflect all economic trade-offs among groups defined by U. If U weights a particular group more heavily at the margin, then optimal input subsidy rates are designed to induce two impacts with respect to that group, other things being equal: an increase in expected welfare and a decrease in risk. There may also be trade-offs at the margin between these two impacts. For example, the larger the marginal disutility of environmental damages relative to the marginal utility of producer surplus, the more the subsidies provide incentives to reduce the use of pollution-increasing inputs and increase the use of pollution-reducing inputs.

There may or may not be trade-offs between expected environmental damages and expected farm income. Depending on the trade-offs among groups as defined by *U*, one could envisage scenarios in which green payments reduce expected environmental damages and simultaneously increase expected farm income. The costs of green payments in this scenario would be borne by other groups, such as consumers, factor suppliers or taxpayers.

Theoretically, optimal green payments would be farm-specific, reflecting the contribution of a specific farm to environmental damages and the weight assigned to a specific farm in the social objective function *U*. Theoretically, optimal green payments would also be applied to all inputs, because in general every input has some impact at the margin (either positive or negative) on pollution. However, both of these conditions are impractical. Budget limitations and transactions costs would limit who was paid under a green payments scheme, what actions would be monitored for compliance and payment, how programmes would address producer heterogeneity, and how much information would be obtained and utilized for policy design and implementation. Optimal site-specific input subsidies might also provide arbitrage incentives that could undermine the system (Shortle *et al.*, 1998).

This raises important issues in the design of second-best green payment schemes that have yet to be addressed. For example, cost-effectiveness may be highly sensitive to how payments are targeted to induce pollution control efforts across critical watersheds and land uses. Cost-effectiveness will be poor if producers who have limited impacts on expected damages are given more weight in the social objective function than those with larger impacts on expected damages. Alternatively, subsidy rates could be applied uniformly across producers in a region to reduce transactions costs and to limit arbitrage opportunities. Uniformity reduces cost-effectiveness because it eliminates opportunities to reduce costs further by targeting producers according to their relative impacts on ambient pollution.

There are also international trade issues to consider. The Uruguay Round Agreement on Agriculture (URAA) calls on participating nations to reduce agricultural programmes and policies that support domestic agricultural production or distort agricultural trade. Agri-environmental policies fall into the permitted set of 'green box' policies in so far as they have minimal distorting impacts on production and trade (Vasavada and Warmerdam, 1998). Strictly speaking, the only green payments that would not alter production are subsidies on 'pure abatement' activities. Such a programme would probably not be very cost-effective and might or might not produce significant environmental gains. A cost-effective plan would alter production practices and land use choices, inevitably impacting production and trade. This would pass muster under the URAA only to the extent that these changes did not negatively impact other countries to the point where they would have legitimate cause for complaint to the World Trade Organization. Green payments outside the terms of the URAA might be possible as part of an internationally negotiated package of agricultural policy reforms involving reductions in other agricultural price and income supports.

Conservation Compliance

Instead of offering farmers payments to adopt alternative practices, existing farm programme benefits could be withheld unless the change was made. So-called compliance mechanisms tie receipt of benefits from unrelated programmes to some level of environmental performance. Examples include the US Conservation Compliance programme to reduce soil erosion and the US Swampbuster programme to discourage the drainage of wetlands (USDA ERS, 1994). As applied to agricultural non-point-source pollution, programme benefits could be withheld if a conservation or water quality plan containing the appropriate technologies was not developed and implemented. Producers would have an incentive to develop the plan as long as the expected programme benefits outweighed the costs of implementing the plan.

The effectiveness of conservation compliance for controlling non-point source pollution is limited by the extent to which those receiving programme benefits are contributing to water quality problems (Abler and Shortle, 1989). For example, US farm commodity programme benefits are concentrated among medium to large cash grain farms in the Corn Belt and Great Plains (Ghelfi *et al.*, 2000). While the impact of agriculture on water quality is certainly an important issue in parts of the Corn Belt and Great Plains, there are many other parts of the US where it is also a critical issue, such as the Chesapeake Bay Region, Gulf of Mexico and Great Lakes Region (USDA ERS, 1997; Kellogg *et al.*, 1997). Participation in US farm commodity programmes is much lower in these regions.

There can also be differences over time in farm programme participation versus water quality impacts. A farm's decision to participate in a farm commodity programme is not based on the social costs of water pollution caused by its activities. Rather, it is based on the net gain to the farmer from participation. During periods of high market prices, farmers have incentives to apply more fertilizers and pesticides in order to raise yields. During these periods, then, the incentives to engage in environmentally damaging activities are greatest. Yet it is precisely during periods of high prices that the incentives to participate in farm commodity programmes are lowest. During the commodity price boom of the mid-1970s, for example, market prices for wheat and feed grains were well in excess of US loan rates. As a result, participation in the programmes for these crops was negligible.

The US experience with conservation compliance

The US Food Security Act of 1985 enacted conservation compliance provisions for the purpose of reducing soil erosion. The provisions require producers of so-called 'programme' crops (wheat, feed grains, cotton, rice) who farm highly erodible land (HEL) to implement a soil conservation plan. Reducing soil erosion has implications for water quality. Violation of the plan would result in the loss of price support, loan rate, disaster relief, CRP and Farmers Home Administration (FmHA) benefits. A recent review (USDA NRCS, 1996) determined that only 3% of the nearly 2.7 million fields required to have a conservation compliance plan were not in compliance. The US Department of Agriculture (USDA) estimates that nearly 95% have an approved conservation system in place. An additional 3.8% are following an approved conservation plan with a variance granted on the basis of hardship, climate or determination of minimal effect.

Evaluations of conservation compliance report minimal or moderate increases in crop production costs and significant reductions in soil erosion (Dicks, 1986; Thompson *et al.*, 1989), though regional assessments show significant variation in costs and benefits. Two studies concluded that conservation compliance is a win–win situation, with increased farm income and reduced soil loss (Osborn and Setia, 1988; Prato and Wu, 1991). However, others show reductions in soil loss achieved only with decreases in net farm income (Nelson and Seitz, 1979; Hickman *et al.*, 1989; Richardson *et al.*, 1989; Hoag and Holloway, 1991; Lee *et al.*, 1991; Young *et al.*, 1991). The majority of HEL can apparently be brought into compliance without a significant economic burden. A national survey of producers subject to compliance found that 73% did not expect compliance to decrease their earnings (Esseks and Kraft, 1993).

Conservation compliance has resulted in significant reductions in soil erosion. Average soil erosion rates on over 50 million HEL acres have been reduced to 'T', or the rate at which soil can erode without harming the long-term productivity of the soil. If conservation plans were fully applied on all HEL acreage, the average annual soil erosion rate would drop from 16.8 to 5.8 tons per acre (USDA NRCS, 1996).

Conservation compliance has been calculated to result in a large social dividend, primarily due to off-site benefits. An evaluation using 1994 HEL data indicated that the national benefit/cost ratio for compliance was greater than 2:1 (although the ratios varied widely across regions) (USDA ERS, 1994). In other words, the monetary benefits associated with air/water quality and productivity outweighed the costs to government and producers by at least 2 to 1. Average annual water quality benefits from conservation compliance were estimated to be about US$13.80 per acre (USDA ERS, 1994).

However, conservation compliance is still limited in what it can accomplish, because it is tied to participation in farm commodity programmes. The conservation compliance's water quality benefits, like farm programme participation, are concentrated geographically in the Corn Belt and Great Plains (USDA ERS, 1997). This means that the programme cannot adequately address water quality concerns in other regions where participation in farm commodity programmes is lower, such as the Chesapeake Bay Region, Gulf of Mexico and Great Lakes Region.

Endnotes

1. This chapter borrows extensively from our previous work – in particular, Ribaudo *et al.* (1999), Ribaudo and Horan (1999), Horan *et al.* (1999a) and Shortle and Abler (1997, 1999). The views expressed are those of the authors and do not necessarily reflect the views of the US Department of Agriculture.
2. We do not compare the relative efficiency of voluntary versus involuntary policy approaches here. Instead, see Wu and Babcock (1999) for an interesting discussion of factors influencing relative performance in the case of agricultural pollution.
3. Producers may also have limited knowledge about the set of alternative production technologies that are available and their economic and environmental characteristics, as well as uncertainty about how their actions impact water quality. For simplicity, these last two situations are not represented explicitly in Fig. 3.1; however, they would have obvious representations.
4. The authority may also have better information about alternative technologies with which the producer is unfamiliar (which could also be represented by T1), and about the relationship between input use and water quality (curve R1).
5. For simplicity, we ignore short-run influences such as risk and learning. Instead, we take a long-run view and assume that a practice will eventually be adopted if education can convince producers that it will make them better off (increase expected utility). We note, however, that uncertainty and other factors could slow or prevent the adoption of practices that might, in the long run, increase producers' net returns while also improving water quality. Such factors represent additional limitations that educational programmes would have to overcome.

Chapter 4

Estimating Benefits and Costs of Pollution Control Policies

MARC RIBAUDO[1] AND JAMES S. SHORTLE[2]

[1]*USDA Economic Research Service, Washington, DC 20036-5831, USA;*
[2]*Department of Agricultural Economics and Rural Sociology,*
Pennsylvania State University, University Park, PA 16802, USA

Water quality degradation imposes economic costs in a variety of forms. The economic benefit of pollution controls is the reduction in these damage costs. Control costs include expenditures by firms on pollution control practices and equipment, increased costs of goods for consumers, and government expenditures on monitoring and enforcement of pollution control policies. This chapter provides an introduction to methods and issues in obtaining valid and reliable estimates.[1]

Basic Concepts and Procedures

There are three basic steps in a prototypical analysis of the economic benefits and costs of an environmental policy (Hanley and Spash, 1993; Hanley, 1999). The first is to identify and forecast the impacts of the policy that give rise to economic costs and benefits, with impacts defined as the difference between the 'state of the world' with the policy versus without. A list of variables that affect economic costs and benefits would include prices, wages, profits, the use of land and other resources, commodity production, environmental quality levels, and direct and indirect human uses of the environment. The next steps are to monetize and aggregate impacts over individuals and time. Studies that follow this plan using state-of-the-art methods are rare. Still, it is worth reviewing some elements of a comprehensive study.

Environmental policy impacts

Environmental policy impacts are generated through a chain of actions and reactions. Some of these involve interactions within the economy, some involve interactions within the environment, and some involve interactions between the environment and economy (Fig. 4.1). In standard analysis, the first reaction is that of producers to the implementation of the policy instrument. Typically, it is assumed that producers take actions to minimize the costs of compliance.[2] Possibilities for individual farmers include going out of business, relocating the farm business, changes in the mix of outputs produced, and changes in production and pollution abatement practices. Given adjustment costs and learning by doing about new practices, the decisions made in the short run will generally differ from those made in the intermediate to long run. Moreover (as discussed in Chapter 3), environmental policies may induce technical change that may bring about further changes over the long run in the technological structure and location of agricultural produc-

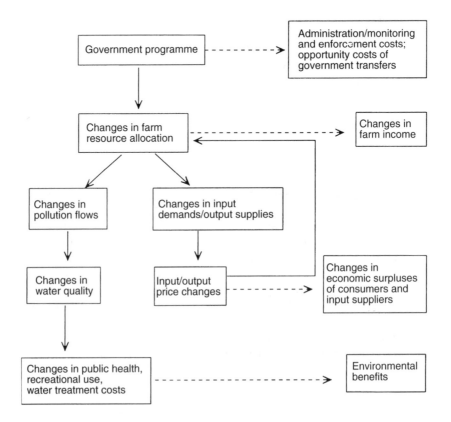

Fig. 4.1. Benefit–cost linkages.

tion. A fundamental contribution of economics to environmental policy analysis is describing and forecasting such behavioural responses to policy initiatives. These forecasts play two crucial roles. One is to provide information for assessing the costs of changes in farm resource allocation. The second is to provide information on changes in agricultural production practices needed for forecasting changes in pollution loads from agricultural lands.

Additional impacts with economic welfare consequences may be generated through economic linkages to input and output markets. If the number of hectares (or producers) directly affected is sufficiently large, then the prices of inputs (e.g. local farm labour or farmland) or farm products may change as a result of change in farmers' demand for inputs. Increased (decreased) input prices would mean benefits (costs) to input suppliers, and would modify the costs in the target farm population. Increased (decreased) output prices would mean costs (benefits) for consumers, and would also modify the costs in the target farm population. Moreover, changes in input or output prices would lead to benefits or costs to farmers who are not directly affected by the environmental measures.

Environmental benefits are initiated by impacts on the volume and timing of pollution flows that result from changes in farm resource allocation. These in turn result in changes in chemical and biological attributes of water resources. Forecasting the impacts of changes in farm practices on environmental quality attributes requires the use of physical models linking farming practices to pollution loads, and to chemical and biological indicators of water quality. Suites of models have been developed for examining the relationships between farm practices and pollution loads, and between pollution loads and chemical indicators of water quality (Novotny and Olem, 1994). These models are the empirical forms of the notional 'runoff' and 'fate and transport' presented in Chapter 5. Factors included in these models are illustrated in Fig. 4.2.

Watershed or basin models can be used to predict how farming practices affect residual loads to receiving waters. While watershed models generally require a great deal of data, depending on the size of the watershed and the complexity of the water system, recent developments in computer hardware and software have allowed for large area simulations of hydrologic processes to be more easily accomplished. An example of a watershed model is the Soil and Water Assessment Tool (SWAT) (Arnold *et al.*, 1995). SWAT is now used extensively in the US. Continued development of databases will allow models such as SWAT to be used more easily in a policy setting.

Another model that is finding increased use in the US is the US Geological Survey's SPARROW model (for Spatially-Referenced Regression on Watershed Attributes). SPARROW is a statistical model that relates stream-nutrient loads to upstream sources and land-surface characteristics (Preston and Brakebill, 1999). It has been used in the assessment of gulf hypoxia (Alexander *et al.*, 2000) and in evaluating policy options for reducing nutrients to the Chesapeake Bay (Preston and Brakebill, 1999).

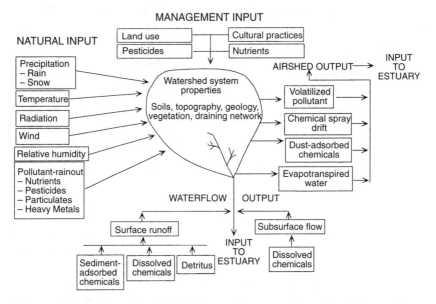

Fig. 4.2. Factors influencing the behaviour and export of agricultural chemicals from an agricultural watershed. (Adapted from Bailey and Swank, 1983.)

Another model that is being used in the assessment of nutrient reduction strategies in the Chesapeake Bay is the Hydrologic Simulation Program – FORTRAN (HSPF) (Donigian *et al.*, 1994). The HSPF model allows the simulation of nutrient loading on the basis of information collected in the watershed. Some other models that have been used in watershed studies include Agricultural Nonpoint Source model (AGNPS) (Young *et al.*, 1987) and Simulator for Water Resources in Rural Basins model (SWRRB) (Williams *et al.*, 1985).

Field-scale models provide the ability to develop and evaluate farm management strategies and policy instruments at a smaller and more detailed scale. Field-scale models generally represent a homogeneous land use, and are used to evaluate on-site performance of best management practices in terms of nutrient and pesticide leaching below the bottom of the crop's root zone and surface runoff of chemicals and sediment past the edge of the farm field. Some popular field-scale models include the Universal Soil Loss Equation (USLE) (Wischmeier and Smith, 1978), Chemicals, Runoff, and Erosion from Agricultural Management Systems model (CREAMS) (Knisel, 1980), Groundwater Loading Effects of Agricultural Management Systems model (GLEAMS) (Leonard *et al.*, 1987), Erosion-Productivity Impact Calculator model (EPIC) (Williams *et al.*, 1984), Pesticide Root Zone Model (PRZM) (Carsel *et al.*, 1984) and Nitrogen Leaching and Economic Analysis Package model (NLEAP) (Shaffer *et al.*, 1991). While such models cannot be used to estimate changes in pollutant loadings to water resources, they can be used to compare alternative policies in their ability to reduce pollutant loss from the field in a least-cost manner.

The USLE is widely used by the US Department of Agriculture (USDA) to estimate reductions in soil erosion from implementing conservation practices, and is used by USDA to enforce conservation compliance on highly erodible cropland. EPIC estimates chemical loss from a field to surface water, groundwater and the atmosphere. It has been built into the USDA's USMP agricultural sector model, enabling a direct link between a policy's welfare impacts to producers and consumers and changes in the generation of pollutants (House *et al.*, 1999).

Chemical indicators provided by watershed or field-scale models (e.g. total suspended solids, dissolved oxygen, total nitrogen, acidity, etc.) have long been standard measures of water quality but, increasingly, scientists are advocating the use of biological indicators (e.g. the presence or relative abundance of indicator species; taxa richness) because trends in chemical indicators can be misleading. For example, standard chemical measures may be improving even while biological conditions are on the decline (Karr and Chu, 1999). Figure 4.3 is a conceptual model adapted from the US Environmental Protection Agency's Waquoit Bay watershed risk assessment (US EPA, 1996). It illustrates relationships between pollution loads, ecological impacts, and assessment impacts and measures. At the top of the model, human activities that cause environmental stresses in the watershed are shown in rectangles. These sources of stressors are linked to stressor types, depicted in ovals. Multiple types of stressor source are shown to contribute to an individual stressor. The stressors then lead to multiple ecological effects, depicted again in rectangles. Some rectangles are double-lined to indicate effects that can be directly measured for data analysis. Finally, the effects are linked to particular assessment endpoints. The connections show that one effect can result in changes in many assessment endpoints. A weak link in current assessment capacity is quantitative modelling of changes in biological endpoints in response to changes in stressors.

Table 4.1 illustrates types of benefits for freshwater quality changes, using a taxonomy developed by Mitchell and Carson (1989). To understand fully the impacts of water quality changes, and properly assess the types of benefits of water quality improvements illustrated in Table 4.1, additional economic modelling is needed. Specifically, changes in water quality will lead to changes in the use of water resources, and other economic responses that influence the benefits of water quality improvements (Smith, 1997). For example, improved freshwater quality in a lake that improves fishing conditions will enhance the experience of current anglers and may lead to an increase in their use of the resource. It may also lead to use of the resource by individuals who had been selecting other sites. These behavioural responses influence the benefits of the water quality improvement. Similarly, an improvement in the conditions of a commercial fishery can affect economic

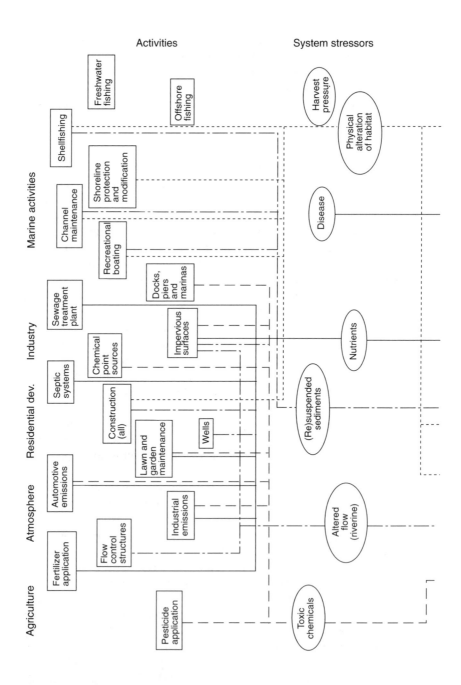

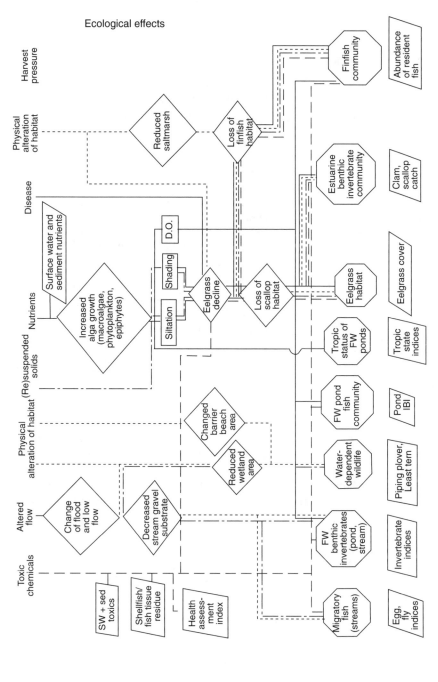

Fig. 4.3. Conceptual framework for watershed risk assessment. (Source: USEPA, 1996.)

Table 4.1. A typology of possible benefits from an improvement in freshwater quality.

Benefit class	Benefit category	Benefit subcategory (examples)
Use		Recreational (fishing, swimming, boating)
	In-stream	Commercial (fishing, navigation)
		Municipal (drinking water)
		Agriculture (irrigation)
	Withdrawal	Industrial/commercial (process treatment)
		Enhanced near-water recreation (hiking, picnicking)
		Enhanced routine viewing (office/home views)
	Aesthetic	Enhanced recreation support (duck hunting)
	Ecosystem	Enhanced general ecosystem support (food chain)
Existence	Vicarious consumption	Significant others (relatives, close friends)
		Diffuse others (general public)
		Inherent (preserving remote wetlands)
	Stewardship	Bequest (family, future generations)

Source: Mitchell and Carson (1989).

welfare through several channels (Freeman, 1993). Initially, improved stocks may reduce the unit costs of fishing, increase the incomes of those who fish and increase harvests. Consumers would benefit as increased supplies lead to reduced prices. Changes in prices, costs and thus incomes from fishing would influence incentives for entry into or exit from the fishery. The resulting benefits to consumers and producers could be strongly affected by the economic structure of the industry (McConnell and Strand, 1989). Again, behavioural responses to the water quality improvement must be examined to assess the benefits fully. As above, it is generally not the impacts on particular individuals or resulting from the decisions of particular individuals that are of interest. Rather, it is statistical (or probabilistic) outcomes on resource uses that are needed.

Valuation

The valuation problem can be developed in a fairly general way as follows. Consider any individual affected by an agricultural environmental policy, because it affects either the individual's income, the prices paid for market goods, or the flow of environmental services received. Examples would include: people receiving farm income; people purchasing farm products or goods produced using farm products; people purchasing goods for which environmental services are an input; people who receive income from the production of goods for which environmental services are an input; and people who directly consume environmental services. The utility function of the individual is denoted $u(x,q)$ where x is a vector of market consumption goods and q is a vector of environmental quality variables.[3] The vector x is selected by the consumer while q is exogenous. The utility function is continuous, increasing and quasi-concave in both vectors.

The consumer's utility maximization problem is to choose x to maximize utility given q and the budget constraint $px \leq m$, where p is a vector of prices for the consumption goods, and m is income. For the problem at hand, the income might be farm income, income from farm-related businesses, or income from a sector in which water quality is an input (e.g. commercial fisheries). The consumer price vector would include the prices of food and other goods produced using farm output as inputs, and the prices of goods produced with environmental services as an input.

The indirect utility function corresponding to this problem is

$$V(p, q, m) = \max_{x}\{u(x, q): px \leq m\}.$$

This function expresses the individual's utility as a function of the prices paid for market consumption goods, income and environmental services. Changes in these variables are therefore the sources of economic benefits and costs and the policy relevant, at least for benefit/cost analysis, impacts.

Let p^0, q^0 and m^0 denote the equilibrium values of p, q and m prior to an environmental policy intervention to reduce agriculture's contribution to water pollution. The post intervention values are p^1, q^1 and m^1. For simplicity, we assume that $q^1 \geq q^0$. Thus the intervention does not reduce the level of any environmental quality variable and increases at least one.[3] Accordingly, if none of the other exogenous determinants of the consumer's welfare change, the consumer will be better off. We impose no structure on the changes in the other variables. It is possible that the environmental policy change may increase the price of some goods, the most obvious possibility being food, and reduce the price of others. Similarly, depending on the type of policy instrument (e.g. taxes, subsidies, regulations) and input and output prices, farm-related income may increase or decrease.

The Hicksian compensating measure of the economic benefit (cost) of the intervention, which we denote as b, is the amount of income that could be taken away from the consumer after the intervention such that utility after the intervention is the same as before. This amount is positive if the policy intervention increases the individual's welfare (e.g. an individual who enjoys improved environmental quality and suffers no adverse impacts on prices or income) and thus a measure of benefit. It is the maximum amount the consumer would be willing to pay (WTP) for the intervention. Alternatively, if the policy intervention reduces the individual's welfare (e.g. an individual who suffers from increased prices or reduced income), the Hicksian measure is negative and thus a measure of cost. The absolute value of the Hicksian measure in this case is the minimum amount the individual would be willing to accept (WTA) for the intervention.[4] Formally, we have

$$V(p^0,m^0,q^0) = V(p^1,m^1 - b,q^1)$$

where b = WTP if b is positive and $-b$ = WTA if b is negative.

Dual to the utility maximization problem is the problem of minimizing the expenditures required to attain a specified level of utility. The expenditure function corresponding to this problem is:

$$e(p,q,u) = \min_x \{px: u(x,q) \geq u\}.$$

Using the expenditure function, b can be expressed as:

$$b = e(p^0,q^0,u^0) - e(p^1,q^1,u^0) + (m^1 - m^0) \tag{1}$$

where u^0 is the pre-intervention level of utility. Accordingly, an individual's benefit (cost) is the reduction (increase) in income required to achieve the initial utility level, plus the actual change in income. Given $m^0 = e(p^0,q^0,u^0)$, we can also write (1) as:

$$b = m^0 - e(p^1,q^1,u^0) + (m^1 - m^0). \tag{2}$$

We can also express b as the line integral (Boadway and Bruce, 1984; Johansson, 1987):

$$b = -\left[\sum_i \int_{p_i^0}^{p_i^1} h_i\left(p,q,u^0\right) dp_i + \sum_j \int_{q_j^0}^{q_j^1} \frac{\partial e}{\partial q_j} \partial q_j \right] + \left(m^1 - m^0\right) \tag{3}$$

where p_i is the ith element of p, $h_i(p,q,u)$ is the Hicksian demand for consumer good i, and q_j is the jth element of q. The integrals under the first summation are the changes in the Hicksian *compensating variation* for each good due to commodity price changes. These terms vanish if prices are unaffected. The integral in the second summation gives the Hicksian *compensating surplus* of the increase in environmental services. Thus, the benefit (cost) of the policy for the individual is given by the sum of changes in the Hicksian compensating variation resulting from changes in the prices of market goods, the sum of the Hicksian compensating surplus due to changes in the levels of environmental services, and the change in income.[5]

Methods for Estimating the Benefits of Water Quality Improvements

Benefits from water quality improvements may accrue because the productivity of goods that use water quality or water quality-related environmental services is increased, or because individuals derive an increase in the welfare directly from the increase in environmental services.

The benefits of productivity improvement

When water quality is a factor in the production of a market good, there are two avenues through which benefits from water quality improvements can be obtained: (i) through changes in the price of the marketable good to consumers; and (ii) through changes in incomes received by resource owners.

Firstly, consider the case of a single firm in a competitive industry. The firm's variable cost function is $c(x,w,q)$ where x is output, w is an input price vector, and q is water quality. Let p be the output price. Water quality increases productivity, reducing production costs. Accordingly, we have $\partial c/\partial q \leq 0$. Profit is given by:

$$\pi = px - c(x,w,q) - FC$$

where FC is fixed costs. Given profit maximization, the firm will choose x such that $p = \partial c/\partial x$. Let $x^*(p,w,q)$ denote the optimal choice of x. This quantity is the firm's supply of the good given the market price p, factor price vector w and water quality level q. The firm's profit function is:

$$\pi(p,w,q) = px^*(p,w,q) - c(x^*(p,w, q), w,q) - FC.$$

Changes in q will affect the firm's cost and profits, and thus the income resulting from production. An exact measure of the income change is the change in quasi-rents, where quasi-rents are revenues less variable costs (Just *et al.*, 1982). Since fixed costs are fixed, the change in quasi-rents (Δ) due to a change in water quality, and any related induced price change, is equal to the change in profits. Suppose the initial economic condition $p = p^0$, $w = w^0$, $q = q^0$. After the water quality change, the new economic state is $p = p^0$, $w = w^0$, $q = q^1$ (for simplicity we have assumed product and factor prices to be unaffected). The change in quasi-rents is:

$$\Delta = \pi(p^0,w^0,q^1) - \pi(p^0,w^0,q^0).$$

By the fundamental theorem of calculus, and using the envelope theorem, we can express Δ as:

$$\Delta = -\int_{q^0}^{q^1} \frac{\partial c}{\partial q} dq.$$

The term $-\partial c/\partial q$ gives the cost savings to the firm at the margin from an increase in q and can be interpreted as the firm's marginal willingness to pay, or marginal benefit, from a water quality improvement. Our second expression for Δ expresses the total benefit as the integral of the marginal benefit for the discrete change.

Alternatively, the benefit can be measured by a change in the firm's producer surplus. This is illustrated in Fig. 4.4, where MC_0 is the marginal cost curve of the firm ($\partial c/\partial x$) given $q = q_0$. Under the assumption of profit maximization, the producer chooses output to equate the price of the good (p_0) with the marginal cost of production, assuming price exceeds the minimum of the average variable costs. Prior to an improvement in environmental quality, output is x_0. Subtracting variable costs, obtained by integrating marginal cost to the output level x_0, from revenues, given by the product of p_0 and x_0, we obtain quasi-rent as the area a. This area is also the firm's producer surplus, defined as the area below the price above marginal cost at the profit-maximizing output.

Suppose a water quality improvement increases q to q_1, shifting the marginal cost curve to MC_1. Assuming that output and input market prices are unaffected, the profit maximizing output increases to x_1. The producer surplus or quasi-rent is now the area $a + b$. The change in producer surplus is the area b. This is a measure of economic benefits to the firm from an improvement in environmental quality when quality enters directly into the production function of the firm.

Holding input and factor prices constant, as we have done in the analysis of the firm, is plausible if the number (or size) of firms affected by the environmental change is small enough to have no impact on prices. If, however, prices are affected, then the firm's welfare effects will be modified by the price changes. Moreover, the welfare of consumers and input suppliers may be affected. We consider the first of these cases.

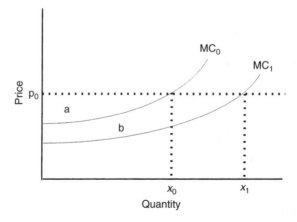

Fig. 4.4. Benefits from an improvement in environmental quality to a single firm.

In Fig. 4.5, let D be demand for good x. Let S_0 be the industry supply curve for x. In market equilibrium prior to the water quality improvement, x_0 is produced and is consumed at price p_0. Assume that an improvement in environmental quality affects firms as shown in Fig. 4.4. As a result, the industry supply curve shifts to S_1, reflecting lower production costs. The equilibrium price of the good falls to p_1, and industry output rises to x_1. The change in producer surplus is the area $A + B - E$. Unlike the case with fixed prices, the change in production surplus may be negative. This would be the case if the producer surplus loss due to the price reduction exceeded the gain from improved productivity.

Turning to the consumer welfare impact of the price reduction, we noted above that economic theory calls for the use of the Hicksian compensating variation. In practice, benefits or costs related to consumer price changes are often approximated by the area beneath the ordinary or Marshallian demand curve (Willig, 1976; Just *et al.*, 1982; Freeman, 1993). Using this approximation, the consumer surplus after the water quality improvement is the area $C + D + E + F$, implying a change in consumer surplus of $C + D + E$. The social benefit from the improvement in environmental quality is the sum of changes to producer and consumer surpluses, or the area $A + B + C + D$. Note the importance of considering the impacts on consumers. Looking only at the industry costs would underestimate the social benefits of the water quality improvement.

For the simple competitive equilibrium model outlined above, estimates of consumer and producer welfare impacts due to productivity enhancing changes in water quality would require knowledge of the information contained in Fig. 4.5. This information would include the impacts of water quality on production costs, the supply of output and the demand for the good. The measurement of benefits becomes more complicated as the number of

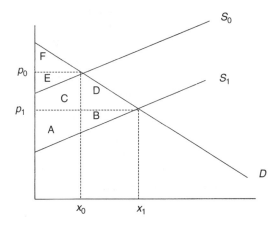

Fig. 4.5. Benefits to consumers and producers in an industry from an improvement in water quality.

markets affected and the complexity of the structure of production, markets and information are increased. Complications include cases in which firms benefiting from productivity improvements produce multiple products and operate in several markets, participate in a vertically linked set of markets, imperfect markets, or markets that are distorted by government policies, experience capital adjustment costs when responding to environment change, make production decisions under uncertainty about economic and environmental conditions, or produce from natural resource stocks (Just *et al.*, 1982; Adams and Crocker, 1991; Freeman, 1993).

Adams and Crocker (1991) divided methods for estimating the benefits of productivity changes into positive methods and normative methods. With normative methods the economic researchers place themselves in the shoes of economic agents, acquire the technical and price data that the idealized agent would use, and solve the economic decision problems. Mathematical programming and simulation models of production and markets are commonly used in normative studies. An example is the McGuikin and Young (1981) study of the costs of desalination of household water supplies. Positive models use data on observed choices of economic agents to estimate econometric models that can be used for analysing impacts and values. An example is the Kahn and Kemp (1985) study of pollution impacts on Chesapeake Bay fisheries. The distinction between normative and positive is not always sharp since some studies blend techniques (Adams and Crocker, 1991; Freeman, 1993). Some of the more interesting and exemplary applications of these techniques involve agricultural applications, but with agriculture as a receptor (particularly of air pollution and climate change) rather than an agent of environmental change (Adams and Crocker, 1991). See Adams and Crocker (1991) for further discussion of the techniques and their merits for benefits estimation.

There are two special circumstances where extensive information on demand for, and supply of, a good are not required. The first is when water quality is a perfect substitute for a purchased input. In this case, improvement in quality results in a decrease of the purchased input. When the change in total cost does not affect marginal cost and output, the cost saving is a true measure of the benefits of the change in quality (Freeman, 1979b). An example could be the reduction in chlorine needed to treat water for drinking as ambient bacterial levels are reduced.

When quality is not a perfect substitute, benefits can sometimes be measured by the change in net returns. If the firm is small relative to the output and factor markets, it can be assumed that product and variable factor prices will remain fixed after the change in quality. The increased productivity is expressed as increased profit calculated from firm budget analysis.

Measuring the benefit of changes in direct consumption of environmental services

The preceding discussion relates to measures (exact or approximate) of the Hicksian compensating variation and income change components of benefits given by equation (3). It remains to discuss measures of the Hicksian compensating surplus component. This is the major challenge in valuing changes in non-market environmental goods, and is the subject of rapidly evolving literature that has produced a suite of approaches.

Although there is no organized market for environmental goods, people do respond to changes in environmental quality. For example, people may alter the number of visits to the recreation sites they visit if environmental quality at one site changes (the demand for recreation changes). When consumers cannot avoid pollution, they may take steps to avert the consequences of poor environmental quality, such as purchasing bottled water if their water supply becomes contaminated with a pesticide. Approaches that are based on changes in observed behaviour are also known as indirect methods (Hanley *et al.*, 1997) and include defensive expenditures, travel cost and hedonic pricing.

In some cases the value a consumer places on an environmental resource is not reflected in observable behaviour. Economists have recognized the possibility that individuals who make no active use of a resource might derive satisfaction from its existence in a particular quality state (Arrow *et al.*, 1993). For example, people may place a value on the existence of a virgin redwood forest and the ecosystem it supports, even if they never plan to visit one. A family of approaches has arisen that uses carefully designed surveys to elicit directly from individuals how they value changes in environmental quality without the need to ascertain how their behaviour might change. These are collectively known as contingent valuation approaches.

Defensive expenditures

For many water quality problems associated with agriculture, a variety of averting or defensive expenditures can be made by individuals to reduce or completely negate the pollution damage. Purchasing water softeners and bottled water are two examples. Change in defensive expenditure has been shown to be a lower bound estimate of benefits from a reduction in pollution (Courant and Porter, 1981; Bartik, 1988a). The theory, from the standpoint of an individual consumer, can be shown using the following arguments presented by Bartik.

The problem for the consumer is to maximize utility,

$$u = u(x,q) \tag{4}$$

with respect to x and q, such that:

$$x + D(q, e) = m,$$

where x = numeraire commodity, q = quality of personal environment, e = pollution level, $D(\ )$ = defensive expenditure function and m = income.

An example of q might be the quality of drinking water in a home, where e is water pollution affecting drinking water, and $D(\ )$ is the cost of water treatment to enhance drinking water quality. The first-order conditions for utility maximization in the choices of x and q reduce to:

$$u_q/u_x = D_q. \tag{5}$$

The household chooses x and q to equate marginal value of environmental quality to the marginal cost of maintaining that level of personal quality.

The benefits from a reduction in pollution are equal to the income required to keep the household at the original level of utility, given the change in pollution. The indirect utility function for this problem can be written as:

$$V = V(e,m) = u(x^*,q^*) + \lambda(m - x^* - D(q^*,e)), \tag{6}$$

where x^* and q^* are the optimal quantities of x and q (given pollution e and income m).

The benefit of a change in p while V, q and x remain fixed is:

$$\left.\frac{\partial m}{\partial e}\right|_v = -V_p/V_m = D_e. \tag{7}$$

The benefit from a small reduction in pollution is D_e, the saving in defensive expenditures needed to maintain the original level of personal environmental quality q^* (and utility) (also shown by Courant and Porter, 1981; Harford, 1984). The results are similar for non-marginal changes in e (Bartik, 1988a).

Actually estimating D_e is not straightforward, since the data requirements for estimating the household demand for personal environmental quality are forbidding (Bartik, 1988a). The observed change in defensive expenditure given an actual change in environmental quality is not equivalent to D_e. Actual change in defensive expenditure can be expressed as:

$$D(Q_0,e_0) - D(q_1,e_1). \tag{8}$$

This measure is an underestimate of true benefits (Freeman, 1979b). In the case of household water treatment, the consumer's desired level of personal environmental quality is higher than previously because of the generally cleaner water, and changes in water treatment activity partially reflect the new goal.

A lower bound estimate of D_e that requires information only on the defensive expenditure function $D(q,e)$ and household choices before and after the pollution reduction is expressed as:

$$D(q_0,e_0) - D(q_0,e_1) \equiv DS. \tag{9}$$

DS is a measure of the change in costs to maintain the initial level of household water quality (not utility, as in the ideal measure D_e). Bartik shows that *DS* is analogous to a Laspeyres measure of the benefits of a price reduction and gives a better estimate than the actual change in defensive expenditures. (*DS* and D_e are exactly the same if the defensive expenditure function is linear.)

Travel cost

Water quality is an important factor in individuals' decisions about many water-based recreation activities. For these 'goods', demand for water quality (a non-market good) can be ascertained through differences in demand for recreation. It is possible to relate variations in quality to changes in demand by making use of the weak complementarity between recreation and site quality (Maler, 1974).

Consider a utility function where utility depends on the consumption of n private market goods (x_i, $i = 1, 2, ..., n$) and water quality (q):

$$u = u(x_1, ..., x_n, q). \tag{10}$$

If there exists a commodity x_i such that u is independent of q if that commodity is not consumed, then that commodity and q are said to be weak complements. This can be shown as:

$$u_q(0, x_2, ..., x_n, q) = 0, \tag{11}$$

where u_q is marginal utility with respect to q. In this expression, x_1 and q are weak complements.

Figure 4.6 is used to illustrate the use of a weak complement to value a change in q. At the base level of quality (q_0), the demand curve for x_1 is D_q, and x_1' is consumed at price p_0. An improvement in environmental quality to q^* shifts demand out to D_q^* and x_1'' is now consumed at price p_0. The area between D_q and D_q^* above the price line, defined by CBDE, is the willingness to pay for the increase in environmental quality, or, equivalently, the area beneath the demand curve for environmental quality (Freeman, 1993).

Recreation sites with water resources as a feature, and water quality at these sites, appear to fit this definition of weak complements. One is indifferent to water quality at that site unless a visit is made (assuming no option or existence value). The travel cost (TC) method is the best known revealed preference technique for recreation valuation in such cases. TC analysis correlates the cost of accessing an outdoor recreation site (the travel cost) with the decision to visit sites. A demand curve is typically generated by regressing the number of visits to the site on travel cost and other exogenous variables (presumably, higher travel costs lead to a diminished visitation, other things being equal). Consumer surplus values can be generated from this demand curve for 'access to the recreation site'.

Consider, for example, the linear travel cost model, where an individual (i) has a demand for trips (T_i) that is modelled as:

$$T_i = \alpha + \beta P_i, \tag{12}$$

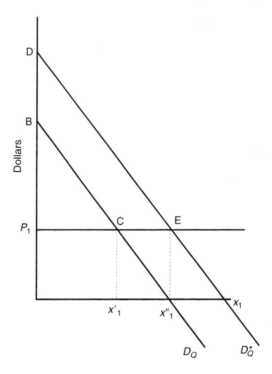

Fig. 4.6. Weak complementarity and revealed preference.

with P_i the individual's travel cost, and β the price coefficient. Integrating between P_i and the cut-off price (the price where demand drops to zero) yields the consumer surplus for recreation at the site, which in this case can be shown to equal $-Y_{i2}/2\beta$.[6]

While originally constructed to provide the full value of a single site, the travel cost model can be extended to recognize the contribution of particular characteristics of a site to individual welfare. The simplest case is when a travel cost study is conducted at a site before and after a change in environmental quality has occurred. The data can be combined, and a site demand function can be estimated that includes environmental quality as a demand shifter (represented as a dummy variable; 0 for the initial level of quality and 1 for the new level). The difference in consumer surplus under the site demand curves for the two levels of environmental quality is the value of the change in environmental quality.

It is possible to estimate the value of water quality without a change having occurred. A number of these extensions are presented below, all based on the insight that if a number of sites exist in a region and environmental quality varies across sites, the demand for quality should be reflected in the relative intensity of use of the sites. Furthermore, with extensive data on the recreation activities of the population, this demand can be quantified and used to ascertain the marginal and infra-marginal contribution of improvements in environmental quality. These extensions of the travel cost model include: (i) site choice models, (ii) hedonic travel cost analysis and (iii) generalized travel cost analysis.

SITE CHOICE MODELS. Site choice models examine the choice of site made by an individual, where at each choice opportunity the individual is faced with many different sites and must choose which site to visit. The individual will presumably choose the site that yields the greatest utility, given the site's environmental quality and cost of access. Information on this 'discrete choice' (to visit or not to visit), when combined with knowledge of the characteristics (including water quality) of the site, can be used to infer the value of these site characteristics (Bockstael *et al.*, 1987). Site choice models are also known as discrete choice models or random utility models.

For example, suppose that on day t an individual faces a set of J sites, and chooses a site j ($j = 1, ..., J$) such that utility (V_t) is maximized:

$$V_t^* = \frac{max}{j \in J} V\left(p_j, q_j, v_{jt}\right) \tag{13}$$

where p_j is the cost of accessing site j (the travel cost), q_j is a vector of characteristics describing the environmental quality at site j, and v_{jt} is a random component that incorporates unobservable factors that influence the individual's enjoyment of site j on day t. Since v_{jt} varies across both site and time, the choice of site is not deterministic. Instead, it is a function of realization of all the v_{jt} ($j = 1, ..., J$).

If the distribution of v_{jt} and the utility function are known, it is possible to compute the marginal value of each element of q_j. For example, it is often assumed that V_t has the following form:

$$V_t = \alpha \, p_j + \beta \, q_j + v_{jt} \tag{14}$$

where α and β are coefficient vectors to be estimated, and v_{jt} is an independent and identically distributed type I extreme valued random variable. In this case, given information on the outcome of many choice opportunities, a multinomial logit model can be used to estimate α and β (Maddala, 1983). Marginal values of site characteristics can then be obtained. An advantage of this approach is that information on the number of visits to each site is not needed, only information on which sites are visited.

Variations of this approach, which generalize the above model, have been widely used (Bockstael *et al.*, 1987; Coyne and Adamowicz, 1992; Adamowicz *et al.*, 1994; Kaoru *et al.*, 1995). For example, Bockstael *et al.* (1987) generalized the distribution of v and added a preliminary stage that predicts the number of choice opportunities. The question of 'the number of choice opportunities' is critical when infra-marginal analysis is attempted, since the discrete choice analysis abstracts from the individual's decision about whether to 'visit a site' or to engage in some other activity, such as staying home and watching TV (Morey *et al.*, 1991).

HEDONIC TRAVEL COST ANALYSIS. In hedonic travel cost analysis, the individual is presumed to derive utility directly from the site characteristics (including environmental quality). In this model, the actual sites are merely particular bundles of site characteristics and otherwise are not unique. If a sufficient number of these 'bundles' exists, demand curves can be identified for each of these characteristics. In other words, for each characteristic, the individual can consume a quantity up to the point where the marginal cost of increasing consumption of the characteristics is greater than its marginal value.

A simple example of a hedonic travel cost model uses a two-stage zonal approach. In the first stage, total trip costs are regressed on several site characteristics to calculate the implicit price of each characteristic. Formally, consider an individual i who has M sites to choose from. For each available site m ($m = 1, ..., M$), the price of access, p_{im} (for example, individual i's travel cost to site m), is regressed against a $K \times 1$ vector of characteristics of the site, q_m: $p_{im} = f(q_m, \epsilon_m)$, where ϵ_m is a random error term.[7] An underlying presumption of this technical (non-behavioural) model of implicit prices is that better sites (say, with increasing water clarity) can be obtained by travelling farther. The derivative of f with respect to q, $\partial p_m / \partial q_k$ can then be used as an implicit hedonic price for an additional unit of each of the (K) components of q.[8]

The second stage requires that the first-stage regression be performed in many different geographical zones (such as counties). Under the likely case that the distribution of sites varies over space, with individuals living in some zones being close to good quality sites while individuals living in other zones are adjacent to lower quality sites, the hedonic price vector for each zone will also vary. An inverse demand curve for each characteristic (k) can then be formed by regressing, across all individuals (i) in all zones, the hedonic prices ($\partial p_m / \partial q_k$) of the characteristic against the quantity of the characteristic demanded (q_{ik}): $\partial p_m / \partial q_k = g(q_{ik})$. These demand curves can then be used for infra-marginal valuation (Englin and Mendelsohn, 1991).[9]

Use of the hedonic travel cost method has brought to light several problems (Smith and Kaoru, 1987; Bockstael *et al.*, 1991). The marginal value of an environmental characteristic, say water quality, is given by the extra costs an individual is willing to pay to enjoy them, i.e. the extra distance the individual is willing to travel to a site with the desired water quality. Yet the location of sites relative to the home of the individual is an accident of nature (Hanley *et al.*, 1997). Sites with desired water quality may be closer to the individual. This has resulted in researchers finding *negative* values for site characteristics which were expected to have positive values.

GENERALIZED TRAVEL COST ANALYSIS. Generalized travel cost starts with simple travel cost models and then correlates the estimated results with measurable site characteristics (Vaughan and Russell, 1982). Thus, environmental quality affects consumers by modifying the price, income and other coefficients of the individual's demand curve for the site. In contrast to hedonic models, it is not postulated that consumers explicitly demand a known level of a site characteristic. In this sense, generalized travel cost models are similar to site choice models. However, unlike site choice models, the total quantity of trips is explicitly modelled, while the choice of 'which site to visit, given we are going to take a trip' is not defined.

Formally, the basic generalized travel cost starts with a set of individual demand curves for $m = 1, ..., M$ sites:

$$
\begin{aligned}
y_{i1} &= \alpha_1 + p_{i1}\beta_1 + \varepsilon_{i1} \\
y_{i2} &= \alpha_2 + p_{i2}\beta_2 + \varepsilon_{i2} \\
y_{iM} &= \alpha_m + p_{im}\beta_m + \varepsilon_{im}
\end{aligned}
\tag{15}
$$

where y_{im} is the observed number of trips taken by individual i to site m, p_{im} is the price (travel cost) of accessing site m, ε_{im} is a random variable and β_m is the price responsiveness for site m (assumed to be the same for all individuals). Each of these individual demand curves is estimated separately, and estimates of β_m, b_m are derived.

The next step is to regress the estimated price coefficient, b_m, against observed site characteristics:

$$
\begin{aligned}
b_1 &= \gamma_0 + z_1\gamma_1 + q_1\gamma_2 + v_1 \\
b_2 &= \gamma_0 + z_2\gamma_1 + q_2\gamma_2 + v_2 \\
b_m &= \gamma_0 + z_m\gamma_1 + q_m\gamma_2 + v_m
\end{aligned}
\tag{16}
$$

where z and q are measures of site characteristics[10] and γ_2 is interpreted as the extent to which price responsiveness changes as these q change. Valuation of changes in q can be generated by computing the price coefficient, b, at the before and after level of q. The difference of the consumer surplus values, with each value derived using a different value of the computed price coefficient (b), will yield an infra-marginal measure of the value of the change in q.[11]

While appealing in its simplicity, the generalized travel cost model suffers from a severe problem: the treatment of substitute sites is not internally consistent. The problem, as pointed out by Mendelsohn and Brown (1983), can be seen by noting that the two stages can be collapsed into a single model with $Y_m = g(p_m, q_m)$, a model that does not include substitute price. For single-site demand curves, exclusion of substitute sites will lead to a problem of missing variables, resulting in a biased estimate of site demand at any given price (Caulkins *et al.*, 1985). For the generalized travel cost models, the consequences are worsened since exclusion of substitute prices is tantamount to assuming that the characteristics of other sites do not affect the demand for a given site. However, if site characteristics are not important in the first stage, characteristics cannot suddenly be the crucial determinant in the second stage. The generalized travel cost technique is basically most useful when applied to a set of unique sites, where each site has no close substitutes and where each site serves a unique market (no individual ever visits two of the sites).[12]

Hedonic pricing

In addition to influencing consumers' decisions on recreational trips, environmental quality can also influence the decision on where to live. For example, houses adjacent to pristine waterways are probably more attractive than otherwise similar houses located next to polluted waterways. The 'amenity value' of a locale's environmental quality can be measured using the property value of homes in the locale (Freeman, 1979a). The underlying assumption is that environmental quality differences tend to be capitalized in land or housing values (Maler, 1977).

Rosen (1974) developed the theoretical underpinnings for what was observed about property values, location and other attributes. The hedonic approach asserts that equilibrium in the housing market can be described by a function relating price to housing attributes, and that the gradient vector of the hedonic function, evaluated at each person's chosen set of attributes, measures the person's marginal willingness to pay for each attribute (Rosen, 1974). Environmental quality can be one of housing's attributes.

The hedonic approach involves two distinct steps. First, the hedonic price equation is used to estimate marginal implicit prices of housing characteristics. Second, marginal implicit prices are used to estimate inverse demand functions or marginal willingness-to-pay functions for groups of households.

The hedonic price equation describes the price of housing (P_h) as a function of its structural (S_{ij}), neighbourhood (N_{ik}), and environmental (Q_{im}) characteristics (Freeman, 1979a):

$$P_{hi} = P_h(S_{il}, ..., S_{ij}, N_{il}, ..., N_{ik}, Q_{il}, ..., Q_{im}).$$

It should be noted that the hedonic approach cannot be used where environmental quality is the same for all residences. There needs to be variation in quality in order for the hedonic price function to be estimated with quality as an explanatory variable. This may be a problem for estimating benefits from changes in water quality on relatively small bodies of water.

The marginal implicit price of a characteristic can be found by differentiating the implicit price function with respect to that characteristic. For an environmental characteristic Q_m, the marginal implicit price is:

$$\partial P_h / \partial Q_m = P_{Qm}(Q_m).$$

The marginal implicit price function gives the increase in expenditure on housing that is required to obtain one more unit of Q_m, ceteris paribus.

If consumers cannot 'arbitrage' attributes by untying and repackaging the bundles of attributes associated with housing parcels, then the hedonic price equation is non-linear, and implicit prices for a household depend on the quantity of the characteristic being purchased (Freeman, 1979b).

Assuming households are price-takers in the housing market and that they select bundles of housing attributes to maximize personal utility, then the implicit price function is a locus of household equilibrium marginal willingnesses to pay (WTP) (Freeman, 1979). A household can be viewed as facing an array of implicit marginal price schedules for the various structural, neighbourhood and environmental attributes. A household maximizes utility by simultaneously moving along each marginal price schedule until marginal WTP for an attribute just equals the marginal implicit price of that attribute. If a household is in equilibrium, the marginal implicit prices actually chosen must be equal to the corresponding marginal WTPs for those characteristics.

Since the marginal implicit price function is a locus of equilibrium household WTP, an additional step is needed to derive household inverse demand curves for attributes from which welfare measures can be derived. The second step of the hedonic technique is estimating the inverse demand function for an attribute by regressing household equilibrium implicit price (w_i) against quantity (Q_{mi}), income (M_i), and other household variables that influence tastes and preferences, or:

$$w_i = w(Q_{mi}, M_i, ...).$$

Household WTP for any level of attribute Q_{mi} can be derived from the inverse demand function.

An important question is whether the demand function for an attribute associated with housing parcels can actually be estimated with information that is generally available and under conditions that characterize housing markets and the prices at which real estate is exchanged. A number of theoretical and empirical problems with the hedonic technique have been identified which raise questions about whether information that can be obtained from real estate markets can be teased to reveal information about the value of environmental quality. One common concern has to do with the real estate market itself. The hedonic approach requires the assumption that each household is in equilibrium with respect to the vector of housing prices and that the vector of prices is the one that just clears the market for a given supply of housing and attributes (Freeman, 1979). If the market for houses is 'thin' or adjusts slowly then observed implicit prices do not accurately measure household WTP (Freeman, 1979; Maler, 1977).

Another concern is whether households can perceive differences in environmental quality (Maler, 1977). This is probably true for some kinds of pollution but not others. If individuals do not perceive environmental quality differences, then property value studies will underestimate the marginal WTP for environmental quality improvements.

Empirical estimation of the hedonic and implicit price functions presents its own problems. These include choice of functional form, the definition of the extent of the market, multicollinearity in explanatory variables, and omitted variable bias (Leggett and Bockstael, 2000). If the price data used to estimate the hedonic model are based on appraisals, tax assessments and self-reporting they may not reflect actual market prices. This poses estimation problems if the error is correlated with other variables (Freeman, 1979).

Even if the hedonic price function can be accurately estimated, it may not provide accurate estimates of welfare changes if discrete rather than marginal changes in environmental quality are being evaluated (Leggett and Bockstael, 2000). Under these circumstances, the hedonic approach may yield only an upper bound for benefits (Bartik, 1988b).

In summary, it is possible for the hedonic model to provide some measure of households' WTP for an environmental amenity (Freeman, 1979; Bartik, 1988b; Leggett and Bockstael, 2000). However, its use is restricted to estimating benefits people experience at or near their place of residence, the underlying assumptions are restrictive, and the empirical analysis must be carefully constructed.

Contingent valuation

Contingent valuation (CV) methods directly ask individuals about their WTP for a general improvement in water quality. They can handle a wide variety of situations, including non-consumptive uses.

The goal of the contingent valuation method is very straightforward: to induce people to reveal directly their WTP for the provision of a non-market good such as environmental quality, or their willingness to accept payment (WTA) to sacrifice the non-market good. This involves asking people, in a survey or experimental setting, to reveal their personal valuations (WTP) for changes in the availability of non-market goods by using contingent, or theoretical, market settings (Randall *et al.*, 1983).

The analyst is interested in evaluating the effect on welfare as a good q (say, environmental quality) changes from a level of q_0 to q_1. A WTP measure, in this case the compensating surplus, can be represented as:

$$\text{WTP}^* = e(p_0, q_0, u_0) - e(p_0, q_1, u_0) \tag{17}$$

where e is the expenditure function, p_0 is the vector of prices for market goods, q_0 and q_1 are the initial and final quantities of the non-market good (environmental quality) and u_0 is utility.[13] When CV studies were first conducted, WTP would often be asked in an open-ended format, along with relevant socio-economic and environmental quality data.

There has been a long and continuing debate as to whether the contingent valuation method actually generates meaningful results. Experience with CV studies (and there have been many) has brought to light a number of troubling problems for the CV approach (Arrow *et al.*, 1993). Economic theory assumes rational choice by consumers. One manifestation of this is that if a consumer considers something to be 'good', then more of that something is better, as long as the consumer is not satiated. This translates into a WTP that is higher for more of a good than less. Some CV studies have reported results that appear to be in conflict with rational choice (Kahneman, 1986; Desvousges et al., 1992; Diamond et al., 1992; Kahneman and Knetsch, 1992).

When consumers make purchasing choices, they face a constraint on their budget. Evidence from CV studies suggests that consumers do not fully consider their budget constraint when they respond to a CV question about a single good when there are other environmental causes they are also likely to support (Samples et al., 1986; Arrow et al., 1993; Hoehn and Loomis, 1993). A piecemeal resource-by-resource approach will overestimate economic value because it does not address substitution possibilities, and respondents are unlikely to consider them on their own. The sum of WTP values for five goods in five separate studies will exceed the total value for the five goods estimated from a single survey (Hanley et al., 1997).

In a similar vein, some studies found that respondents reconsidered their answers when specifically asked to consider what they would give up to make the 'promised' payment, or when asked actually to contribute to the cause they said they were willing to support (Seip and Strand, 1990; Duffield and Patterson, 1991; Kemp and Maxwell, 1992). This can be taken as evidence that consumers are not fully considering their WTP in the context of other, day-to-day needs that are supported by their incomes.

In order for a CV study to collect useful information, the respondent must understand exactly what is being asked, and must accept that information. The use of CV for non-use goods poses a special problem, because consumers have little or no experience with the good. If an exacting description of the problem, consequences of failing to act, and the time and effectiveness of protection are not provided, a respondent will probably end up valuing something different than the surveyor intended.

Even if respondents are provided with detailed information, there is no guarantee that they will accept it all. They may not agree that the consequences are as great as claimed, or that the government could actually provide the protection promised. Again, the end result is that the respondent values a good different than what the researcher intended.

Since CV is a survey technique, identifying the correct population to survey becomes exceedingly important. Non-use goods pose a special problem because the extent of the market is not obvious. How large is the potential market for the Grand Canyon? For an endangered species? For a local historic landmark?

Some critics of CV have questioned whether values reflect actual WTP for the good in question, or a sign of support for environmental protection in general (Arrow et al., 1993). If the latter is true, then WTP as estimated from a CV survey is an unreliable indicator of value. Evidence of this criticism comes from bi-modal distribution of results sometimes seen for open-ended CV questions: responses clustered around 0 and a sizeable, positive number.

This distribution matches the distributions for donations to charities, where people give to only a few favourites.

A review of the CV technique sponsored by the National Oceanographic and Atmospheric Administration (NOAA) concluded that it is exceedingly difficult to conduct, but is capable of providing useful information if the CV instrument is carefully constructed and the appropriate sample taken (Arrow *et al.*, 1993). The review panel recommended a number of guidelines for conducting CV studies, as follows.

1. Probability sampling. It is essential to identify the appropriate sample population for a particular environmental good, and then to design a sampling strategy that will give statistically relevant results.

2. Minimize non-response (zeros) through probing questions. Non-response is believed to be a sign that respondents do not accept the information provided.

3. Personal interviews are preferred, and mail questionnaires discouraged.

4. Pre-test survey instruments to make sure that adequate information is provided to the respondents and that it is accepted.

5. Report all sampling and response information as part of the study report.

6. Conservative design. When aspects of the survey design and analysis of responses are ambiguous, it is recommended that the options that tend to underestimate WTP be used.

7. Pose questions as WTP rather than WTA. WTA is not bound by income and may be more prone to frivolous responses.

8. Referendum format rather than open-ended. Referendum format asks an individual whether they are willing to pay a given amount to protect a resource. The respondent responds with either a yes or no. Statistical methods exist for deriving WTP from such a survey design. The belief is that the respondent can much more easily give a 'true' response to such a question.

9. Accurate description of programme and policy, including how the resource responds over time to protection efforts.

10. Reminder of existence of substitutes for good in question, as well as the budget constraint.

11. 'No answer' option should be allowed. Respondents who answer this way should be asked indirectly to explain their choice.

12. Collect appropriate socio-economic and attitudinal data.

CHOICE EXPERIMENTS. A relatively new extension of CV that may overcome some of its weaknesses is choice experiments (CE) (Adamowicz *et al.*, 1998; Hanley *et al.*, 1998b). The CE method is based on random utility theory and employs a series of questions that are designed to elicit a respondent's preferences for environmental attributes. For recreational fishing, attributes might include species type, catch rates and attributes of sites that influence the experience (Heberling *et al.*, 2000). The values of the environmental attributes can be estimated from the stated preferences.

CE is a generalization of CV in that rather than asking a respondent to choose between a base case and an alternative, CE asks a respondent to choose between multiple cases that are described by bundles of attributes. The bundles of attributes comprise specific scenarios that are selected from the universe of possible scenarios. CE therefore requires very careful design in both scenario development and statistical design. The attribute space covered by the scenarios must cover the range of possible outcomes defined by the relevant policy questions. Statistical design theory is used to construct orthoganol choice scenarios that can yield parameter estimates that are not confounded by other factors (Hanley *et al.*, 1998b). Some of the important decisions that must be made in the design stage include the number of attributes, the number of levels to allow each attribute to take, what these levels should be, and how both levels and attributes should be described.

The CE method has several advantages over CV methods. It provides a richer description than CV of the trade-offs individuals are willing to make between attributes (Adamowicz *et al.*, 1998). This is important, because many management decisions change attribute levels rather than result in the creation or loss of the environmental good. CV is better suited to these all-or-nothing scenarios (Hanley *et al.*, 1998a).

The CE method provides the opportunity to value marginal changes in attributes that may be difficult to observe using revealed preference approaches. This provides more opportunities for benefit transfer, in that it is easier to transfer particular attributes, and their values, from one setting to another than the entire environmental good.

Examples of benefits

While there are numerous studies of the benefits of water quality improvements using the techniques outlined above, empirical estimates of the economic benefits from reductions in water quality-impairing agricultural pollutants are very scarce. The difficulties in linking policy to changes in level of water quality-related services and the non-market nature of most benefits have discouraged research in this area. Ribaudo (1989) estimated the water quality benefits from reducing soil erosion on cropland through the USDA Conservation Reserve Program. A variety of methodologies were used, depending on the available data. A travel-cost model was used to estimate benefits to recreational fishing, using data from a national survey of fishing recreation and from water quality monitoring stations. A cost model of the drinking water industry was used to estimate reductions in drinking water treatment costs from reductions in suspended sediment. Total benefits were estimated to range between $2 and $5.5 billion (1986 US dollars).

Recreation benefits from the CRP were further revisited by Feather *et al.* (1999). A combination of a discrete-choice model and a travel-cost model was used to estimate benefits to freshwater recreation (fishing, swimming and boating) from reductions in soil erosion. Annual benefits were estimated to be about $35 million.

Crutchfield *et al.* (1997) used a contingent valuation survey to estimate the benefits from protecting groundwater from nitrate contamination in four US watersheds. Annual benefits from groundwater that met USEPA's health standard were estimated to be about $314 million for the 2.9 million households in the four regions.

Smith (1992) summarized and extended the findings of Ribaudo (1989), Anderson and Rockel (1991), Powell (1991), Whitehead and Blomquist (1991), Loomis *et al.* (1991) and Poe and Bishop (1992) to derive an estimate of the environmental costs of agriculture relative to the value of crops produced. He estimated that environmental costs associated with soil erosion, wetlands conversion and groundwater contamination ranged from less than 1% to over 40% of the value of crops produced per acre on land deemed responsible for these impacts, depending on the region.

Cost Estimation

Pollution control costs include: (i) the costs of resources devoted to pollution abatement, (ii) the costs of changes in the products farmers choose to produce and the production processes and inputs they use as they seek to minimize the costs of compliance, (iii) the costs to consumers, producers and resource suppliers of input and/or output price changes resulting from changes in market demands and suppliers, and (iv) the social costs of monitoring, enforcement and other administrative activities.

The costs of changes in agricultural production

Analogous to the measurement of benefits to firms that enjoy productivity gains from water quality improvements, the costs to farm firms that must comply with pollution control policies are appropriately measured by changes in quasi-rents (Just *et al.*, 1982).[14] If the prices of output or inputs are affected, then welfare costs to consumers of suppliers of farm inputs must also be considered.

To illustrate, consider a group of farmers producing a single good under perfectly competitive conditions. Let $x = f(z)$ be the production function of a representative farm, where x is output and z is an input vector. In the absence of environmental policy, the farm firm maximizes profit, $\pi = px - c(x,w)$, where p is the output price, w is the input price vector, and

$c(x,w) = \min\{wz : f(z) \geq x\}$.

Let p^0 and w^0 denote the pre-policy values of p and w, and π^0 the pre-policy baseline optimized profit. The cost to the firm of a water pollution control instrument is the reduction in the quasi-rent from the baseline level.

As described in Chapters 2 and 3, there are many possible instruments to induce farmers to undertake changes in production practices to reduce pollution. The ways in which they respond, and the impacts on their welfare will vary with the instrument. Input taxes or subsidies discourage the use of polluting inputs (e.g. nutrients or pesticides) and/or increase the use of pollution control inputs and can be easily analysed with the model we have presented above.[15] To illustrate, let t denote a vector of tax/subsidy rates applied to the representative farm's input vector. Since the tax/subsidy is equivalent to a change in input prices, the post-policy profit is $\pi_1 = px^* - c(x^*, w + t)$ where x^* indicates the optimized output. The change in quasi-rent is:

$$\Delta = \pi^1 - \pi^0$$

This change in quasi-rent can be depicted graphically, and measured in principle, as a change in consumer surplus (with the firm being the consumer of inputs) using the farm's input demand curves, or a change in producer surplus using the farm's supply curve (Just *et al.*, 1982). To illustrate, suppose that only one input, for example fertilizer, is taxed, and that market input and output prices are unaffected by the intervention. Let the first element of w be the fertilizer price. Then w_1^0 is the baseline fertilizer price and $w_1^1 = w_1^0 + t$ is the farmer's fertilizer price with the tax, where t is the tax rate. Using standard results of applied welfare theory, Δ can be expressed as (Just *et al.*, 1982):

$$\Delta = -\int_{w_1^0}^{w_1^0+t} z_1\left(p^0, w\right) dw_1$$

where $z_1(p, w)$ is the farm's fertilizer demand function. This result is given by the area a + b in Fig. 4.7. The line D is the input demand curve. z_1^0 is the quantity of fertilizer demanded before the tax and z_1^1 is the quantity demanded with the tax. This farm level analysis can be easily extended to the input market if all demanders are taxed uniformly.

In output space, the fertilizer tax would imply an upward shift in the farm's marginal cost and supply curve, implying a reduction in producer surplus. This is illustrated graphically using Fig. 4.8. MC_0 is the marginal cost curve of the farm. Under the assumption of profit maximization, the producer equates price (p_0) with marginal cost (assuming price exceeds the minimum average variable cost). Prior to the pollution policy, output is x_0. The quasi-rent, or producer surplus, is the area a + b. Analogous to an input price increase, the tax increases production costs, shifting the marginal cost to MC_1. Assuming the market price remains at p_0, output is reduced to x_1. Producer surplus for this level of production is area b. The change in producer surplus is the area a.

As another example, consider a tax, applied at the rate t_r, on a farm level environmental performance proxy that aggregates over inputs (see Chapter 2). The proxy might be an estimate of field losses of pesticides, nutrients or

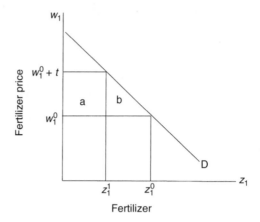

Fig. 4.7. The welfare cost of an input tax.

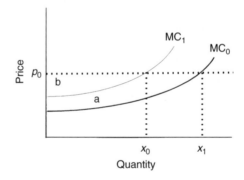

Fig. 4.8. Costs of an environmental tax to a single firm.

soil. Let $r(z)$ give the proxy as a function of the farm's inputs. In contrast to the analysis presented above for input taxes, the proxy tax does not simply increase the prices producers pay for inputs. Accordingly, the cost function must be modified to include the tax as a separate argument. Specifically, the farm's cost function becomes:

$$c_r(x, w, t_r) = \min_z \{wz - t_r r(z) : f(z) \geq x\}$$

The farm's profit function with the tax becomes $px^* - c_r(x^*, w, tr)$ where x^* denotes the optimized output. The change in quasi-rents is:

$$\Delta = \pi^1 - \pi^0$$

By the envelope theorem, the cost to the firm of a marginal increase in the tax on the proxy is $-r(z(p,w,t_r))$, where $z(p,w,t_r)$ is the vector of profit maximizing inputs. Analogous to our treatment of the cost of an input tax, the cost of the tax to the farm firms can be analysed graphically and measured by the change in the producer surpluses.

If the number of producers subject to an environmental policy is sufficiently large to affect the market supplies of agricultural commodities or the market demands for resources that are not perfectly elastic in their supply to agriculture, then the welfare impacts will include price-induced changes in the welfare of producers, consumers and input suppliers. The same techniques we used to illustrate the impacts of an increase in environmental quality on a perfectly competitive industry could be used to analyse the agricultural commodity market impacts of an increase in production costs due to an environmental policy for the simplest case of a sector producing a single commodity. In contrast to the prior analysis, the industry supply curve would shift to the left as a result of the policy-induced cost increases. The market price would rise, reducing consumers' surplus. As in the case of the productivity change, the impact on producers' surplus would be ambiguous since the welfare losses due to the production cost increase will be offset to some degree by the gains resulting from the price increase.

As with the measurement of the benefits of production cost reductions induced by water quality improvements, the measurement of the costs of changes in farm resource allocation for water quality protection becomes more complicated as the complexity of the structure of production, markets and information is increased. Topics include welfare measurement when producers are uncertain about prices, technology, weather and policy, produce multiple products or products that may vary in quality characteristics, experience capital adjustment costs, and participate in input or output markets that are imperfect or distorted by government policies (Just *et al.*, 1982).

Producer–consumer costs versus social costs

The preceding discussion focuses on the costs of water pollution controls to producers and consumers. It is important to note that the costs to society may differ from the costs to producers and consumers. To illustrate, consider again the tax imposed on a farmer's use of fertilizer (Fig. 4.7). The tax increases the unit cost and reduces fertilizer use. Since the farmer was maximizing profits before the tax, the reduction in use implies a reduction in pre-tax profits. The total cost to the farmer is this pre-tax profit reduction plus the profit loss due to the tax payment (area a + b). The social cost is simply the reduction in pre-tax profits. The loss to the farmer from the payment of the tax (area a) is exactly offset by social gain from the use of the tax revenues.

Alternatively, suppose the farmer is paid a subsidy to reduce fertilizer use. Again, if the farmer was maximizing profits before the tax, then the reduction in use implies a reduction in profits before the receipt of the subsidy payment. The cost to the farmer is this profit reduction less the subsidy payment. Depending on how the subsidy is offered, the farmer could well be better off. The social cost is simply the reduction in pre-subsidy profits. The gain to the farmer from the payment of the subsidy is exactly offset by social loss from the use of the public funds.

Finally, consider a policy regulating fertilizer use through the use of permits that limit the quantity that may be used. If the permits are tradeable, the opportunity cost of fertilizer use is the market price of fertilizer plus the market price of fertilizer permits. The increased cost will lead to reduced fertilizer use, and a profit loss before transfers related to permit trades. However, farms that are net permit sellers will receive payments that offset their profit losses, analogous to the payment of subsidies, while farms that are net purchasers will make payments that increase their losses, analogous to taxes. On balance, the monetary transfers between private agents are a social wash.

Social cost measurement with agricultural and fiscal policy distortions

Some important implicit assumptions in our analysis to this point are that farm commodity prices reflect the marginal social value of the goods to consumers, farm input prices represent the marginal social opportunity cost of the resources, and the social opportunity cost of a dollar of tax or subsidy is equal to a dollar. Adjustments in the measurement of social costs of pollution controls may be required if these conditions are not satisfied.[16] There are several important cases to consider.

One set of problems emerges from pre-existing *agricultural policy distortions* (e.g. see Lichtenberg and Zilberman, 1986). Agricultural markets are often subject to policy interventions intended to serve farm income, trade, food price, revenue or other policy goals (see Chapter 3). The interventions take a variety of forms, including output price floors and subsidies, production quotas and input price subsidies (Gardner, 1987). These interventions can affect the location, type and severity of agricultural externalities (Antle and Just, 1991; Hrubovcak *et al.*, 1989; Weinberg and Kling, 1996). Furthermore, they cause divergences in the prices producers pay for resources or receive for their products, and their respective social values.

To illustrate, suppose that producers receive a price subsidy. The market price will then be less than the marginal production cost, implying excess production and a dead weight efficiency loss. An analysis of the costs of agricultural pollution controls must include the effects on the dead weight costs. If a policy reduces (increases) the dead weight cost, then the social cost is reduced (increased) in comparison with the normal measures presented above. The same would be true of environmental policies that affect the deadweight costs of other types of commodity market or input market distortions.

Environmental policies that involve payments to farmers must consider the *social opportunity cost of public funds.* Analogous to our assumption to this point, most studies of the welfare consequences of agricultural policies assume that the opportunity cost of $1 of government spending is $1 (Altson and Hurd, 1990). However, the opportunity cost is generally in excess of $1 because of distortions of labour and other markets resulting from tax mechanisms (e.g. Atkinson and Stiglitz, 1980). The marginal costs vary from country to country, but can be substantial (e.g. Browning, 1976). The implication is that the subsidy payments to reduce polluting inputs or increase the use of pollution control inputs are not appropriately treated as transfers without efficiency consequences. Instead, the social costs of the funds will exceed the value of transfer to the recipients, and thus must be considered. Similarly, the social value of a dollar of environmental tax revenues may be more or less than a dollar depending on the use that is made of the revenues and the other factors (e.g. see Oates, 1995).

Estimating the costs of changes in agricultural production

Because the costs of environmental policies largely involve the welfare impacts of changes in the allocation and prices of goods and services traded in markets, their estimation is generally considered to be less challenging than estimating the benefits. And, in contrast to the limited research on the benefits of reducing agriculture's contribution to water pollution, a wealth of studies estimate the costs of pollution controls in agriculture. A sample of this research is listed in Table 2.2 in Chapter 2.

A simple and common way for estimating pollution control costs is the *direct compliance costs* approach (USEPA, 2000c). Direct compliance costs would include expenditures to install and operate pollution control equipment, and possibly other *partial budgeting* costs. To illustrate, confined animal feeding operations are increasingly subject to regulations on the handling, storage and disposal of animal wastes to reduce odours and water pollution risks. The direct compliance costs of such regulations are the costs of installing and operating facilities to collect and store manure, and the direct expenditures of disposing of manure in accordance with the regulations. In some regions, farmers are subject to taxes on fertilizers. In this case, no expenditures are required for pollution control equipment. A direct compliance cost accounting (or partial budgeting) would estimate the social cost as the pre-tax reduction in net returns due to the reduction in fertilizer use holding the crop mix and other inputs constant. Essentially, this approach would use the marginal value product curves for fertilizer for a given crop mix rather than the fertilizer demand curve to estimate the cost of the tax.

A major limitation of the direct compliance cost method is that it does not allow for input or output substitution effects that producers may choose to reduce compliance costs. For example, in the fertilizer example, depending on the size of the tax and relevant price elasticities, a tax-induced increase in fertilizer costs will lead farmers to substitute other inputs for fertilizer and change their crop mix in favour of less fertilizer intensive crops. Such responses will reduce compliance costs relative to those that would be obtained by the direct compliance method. If new consumer prices are calculated using the direct compliance, they too will be overestimated resulting in an exaggerated loss to consumers, especially if the elasticity of demand is assumed to be zero (i.e. if consumers like producers do not respond to changes in costs) (USEPA, 2000c). None of the studies listed in Table 2.2 use the direct compliance cost approach.

More advanced methods model responses of farm firms and markets. Firm level models are especially useful for detailed analysis of the potential economic responses of farmers to policy initiatives, and the resulting compliance costs and their distribution by structural factors such as farm size, and the impacts of farmers responses on the movement of pollutants from farm fields. Numerous studies can be found at these scales. Larger scale models are required to analyse costs at watershed or other regional scales. Watershed and other regional models imply spatial aggregates of farms and fields. Farms and fields may be modelled individually, but the increased data requirements and computational complexity that attend larger spatial scales typically necessitate the use of less detailed representations of economic and physical variables and relationships than those included in firm level models. Individual economic units will typically be replaced by regional economic aggregates. Possible representations include regional production functions, commodity supply and input demand functions, or cost functions. Representative farms are frequently defined to capture spatial and structural variations in production and pollution control technologies, farm responses to policy initiatives, and costs. For example, the USMP model used by US Department of Agriculture's Economic Research Service is a national model that subdivides the nation into 45 subregions. Each subregion is modelled as a multi-product farm firm. Models with endogenous output or input prices will include commodity demand and farm input supply relationships.

The positive-normative taxonomy of economic modelling approaches presented previously (Adams and Crocker, 1991) applies to models used for analysing policy responses and costs. Among normative techniques, mathematical programming models have found frequent use for modelling firm and sectoral policy responses and compliance costs at all scales. Most of the studies listed in Table 2.2 use this approach. Reasons for the popularity of mathematical programming models include: (i) they can be constructed when data for estimating economic relationships are absent or when available data are inapplicable, which makes them well-suited for *ex ante* policy analysis; (ii) the constraint structure inherent in programming models is well suited to characterizing various forms of resource, environmental and policy constraints; (iii) the optimization nature of the procedures is consistent with economic theory of firms and market behaviour (Adams and Crocker, 1991;

Howitt, 1995). Programming models are not without limitations. A major challenge is to represent the choice technology choice set adequately to capture the range of plausible compliance options.

Positive models, and in particular, econometric models have a strong appeal for environmental policy analysis because they allow results to be inferred from the actual choices of producers and consumers. Their use is, however, generally quite limited for *ex ante* analysis of environmental policy options. Without prior experience, there can be no data on actual choices. This may be of little consequence for options that induce changes in the economic environment that are analogous to past economic experiences. For example, responses to variations in fertilizer and pesticide prices will provide data for the analysis of taxes on these inputs (e.g. Shumway and Chesser, 1994; Wu and Segerson, 1995). However, many of the approaches that are of interest have no analogues.

Policy cost analysis does not require integration of economic and physical models of the formation, transport and fate of pollutants. For example, a study might examine the economic consequences of a nitrogen fertilizer tax on fertilizer use and farm profits without examining the impacts on the movement of nitrogen into water resources. Integration of economic and physical models is essential for examining the water quality impacts of policies, and for assessing the cost-effectiveness of alternative options in achieving water quality goals. While clearly desirable in general, integration of economic and physical models can be challenging, especially as the spatial scale is increased. The studies presented in Table 2.2 provide a range of examples of model integration.

Lessons from policy cost estimates

There are a number of lessons that have been learned from studies of the welfare costs of improving the environmental performance of agriculture. We list some important lessons below.

1. *Watershed-based management.* A major lesson emerging from the research is the importance of watershed- or aquifer-based strategies that simultaneously consider both point and non-point sources. For example, US water pollution control policies have traditionally focused on point sources. There is now substantial evidence that the economic efficiency of surface water pollution controls in the USA could be significantly improved through strategies that allocate load reductions between point and non-point sources based on the relative costs of control (Elmore *et al.*, 1985; Malik *et al.*, 1993; Letson, 1992; Rendleman *et al.*, 1995; Anderson *et al.*, 1997; USEPA, 1998c; Faeth, 2000; GLTN, 2000). (See section on Point/non-point trading in Chapter 2.)

2. *Targeting.* Another important lesson is that targeting agricultural controls in space and time can greatly enhance the cost-effectiveness of pollution reductions (e.g. Braden *et al.*, 1989; Bouzaher *et al.*, 1990; Babcock *et al.*, 1997; Carpentier *et al.*, 1998). Although there is a large degree of uncertainty about the responsibility of individual farms or land areas to non-point pollution loads, there is an increasingly sophisticated scientific understanding of the relationships between land use activities, land characteristics, weather and nutrient losses from farms, and the movement of the spatial transport of nutrients. This information can provide a basis for targeting critical watersheds and land uses (see Chapter 2).

3. *Policy instruments.* A major focus of cost research has been the relative cost-effectiveness of alternative policy instruments for agricultural sources. Both *ex ante* studies of proposed instruments, and *ex post* studies of actual policies demonstrate that choices among compliance measures, and the details of the incentives or regulations, can have a significant impact on the economic and ecological performance. Perhaps the most important lesson from this research is the importance of empirical evaluation of proposed instruments (see Chapter 2).

4. *Win–win opportunities.* Economic analysis of compliance costs generally begins with the assumption that producers maximize their welfare. This assumption implies that the cost of policy interventions must be positive. However, farmers, like other producers, vary in their technical knowledge and management skills. Some may fail to fully optimize. When suboptimal behaviour from a private perspective leads to the use of environmentally harmful production choices, there will be opportunities for win–win improvements. There is research indicating that such win–win opportunities are sometimes available or can become available with technical change (e.g. Shortle *et al.*, 1992; Erwin and Graffy, 1995; Van Dyke *et al.*, 1999), although the literature generally suggests that significant improvements will typically come at a cost (see below). Moreover, while some situations may initially appear to provide win–win opportunities, these opportunities may not always be realized in ways that benefit the environment. The ultimate environmental impacts depend on how farmers take advantage of their ability to optimize, and could reduce environmental quality if, for example, optimization resulted in greater intensification of chemical use (Ribaudo and Horan, 1999; see Chapter 3).

5. *Policy coordination.* As we noted above, a number of studies indicate that price supports, input subsidies and other agricultural policies influence the nature, size and spatial distribution of agricultural externalities. Associated with these policies are deadweight efficiency losses separate from their impacts on water pollution costs. An important implication is that policy reforms can offer some environmental gains at low or possibly even negative social costs (see Chapter 2).

6. *Pollution control is costly.* A consistent finding of the research is that control costs are increasing. While some gains may be achieved at relatively low cost, the costs of stringent controls can be substantial. This point is illustrated by a recent study of the costs of reducing nitrogen loadings to the Gulf of Mexico from the Mississippi River through restrictions on fertilizer use and restoration of wetlands that filter nitrogen. Welfare impacts to consumers and producers for a 20% reduction in fertilizer applications in the Mississippi Basin were estimated to be a loss of $335 million, including environmental benefits from reduced soil erosion (Doering *et al.*, 1999). A 45% reduction in fertilizer applications was estimated to result in a loss of $2.9 billion. Restoring 10 million acres of wetland was estimated to result in a loss of $1.9 billion.

Administration costs

There is comparatively little research on the costs of administering agricultural environmental policies. Recent exceptions are Carpentier *et al.* (1998) and McCann and Easter (1999). Both studies demonstrate significant differences in costs across different types of policies. For the policies examined by Carpentier *et al.*, administrative costs ranged from 4% to 12% of the total costs. The higher percentage was for the policy with the lowest total costs.

Endnotes

1. The standard reference for theory and methods for valuing environmental costs and benefits is Freeman (1993). Other useful sources on applied welfare economics and valuation include Just *et al.* (1982), Johansson (1987) and Hanley and Spash (1993).
2. Experience from other sectors indicates that actions to delay compliance and non-compliance are also to be expected (Russell *et al.*, 1986).
3. The environmental quality vector is inversely related to the ambient concentration variable in the model presented in Chapter 2.
4. The compensating measure of welfare change implicitly assumes that consumers have a right to only their base level of environmental quality. For an improvement in environmental quality, the compensating measure assumes consumers must pay for an improvement in environmental quality. Polluters implicitly have a right to pollute, and they must be compensated for decreasing pollution. For a decrease in environmental quality, the compensating measure assumes that consumers have a right to a cleaner environment, and that polluters must compensate consumers for any degradation.
5. The value of b is independent of the path of price and quantity changes assumed in the evaluation of the line integral. The compensating variations and surpluses associated with particular price and quantity changes are not (Boadway and Bruce, 1984; Johansson, 1987).
6. See Bockstael *et al.* (1990) or Hellerstein (1992) for discussion of consumer surplus calculation in linear demand models.
7. 'Available' means that the site is not dominated by some other site, with domination occurring when one can obtain the same quantity of characteristics from a site which has a lower price. The (zone-specific) available sites are usually proxied by a list of all sites actually visited by individuals from the zone; non-visited sites are assumed to be dominated.

8. If f is linear, then the derivative for each characteristic equals the estimated coefficient. For non-linear hedonic price equations, additional assumptions may be required (Mendelsohn, 1984).

9. As with discrete choice models, hedonic models focus on a single choice occasion and do not explicitly model the number of times an individual chooses to 'visit one of several sites'. If the total number of site visits changes substantially when site quality improves (degrades), these infra-marginal values will be biased downward (upward).

10. Equation 16 differentiates between z and q for heuristic reasons: where z variables may consist of non-environmental variables (say, number of picnic tables in the campground) that may be of lesser interest to the analyst.

11. This simple model can be expanded, such as by modelling the correlation between the error terms (permitting the joint estimation of both stages) or by using censored demand curves (Smith and Desvousges, 1985).

12. An ad hoc method of dealing with this problem is to include a substitute price in a pooled (single) equation version of the generalized travel cost model (Vaughan and Russell, 1982). However, this does not address the fundamental problem of incorporating the characteristics of alternative sites in a demand model.

13. U_0 is the utility obtained at initial prices, income, and q: $V(P_0, Q_0, Y_0)$.

14. This is strictly true provided that production choices are made under conditions of perfect information, the decision environment is static (i.e. no capital adjustment costs), and the firm seeks to maximize profits. Appropriate measures when these assumptions are relaxed are discussed in Just *et al.* (1982).

15. The same would be true of instruments applied to outputs.

16. The issues that we raise here for the measurement of the costs of pollution control may also apply to measurement of productivity benefits.

Chapter 5

Non-point Source Pollution Control Policy in the USA

MARC RIBAUDO

USDA Economic Research Service,
Washington, DC 20036-5831,
USA

Water quality in the US is addressed by literally hundreds of national, state and local laws and programmes using a large suite of policy instruments. Some of these laws and programmes establish water quality goals. Others provide mechanisms for achieving goals. Some apply to agriculture, but many do not.

The focus of this chapter is policies to control agricultural non-point pollution. To provide context, the chapter begins with an examination of water quality problems that drive policy development. An overview of the general water quality policy framework in the US is then presented. The remainder of the chapter addresses agricultural non-point pollution policies.

Water Quality Issues in the US

Aggressive national initiatives to reduce water pollution in the US began with the passage of the Clean Water Act (33 U.S.C. sections 1288, 1329) in 1972. The Act established national water quality goals and provided regulatory means for pursuing them. Since the passage of the Clean Water Act in 1972, water quality in the US has improved largely through reductions in toxic and organic chemical loadings from point sources. Discharges of toxic pollutants have been reduced by an estimated billion pounds per year (Adler, 1994). Rivers affected by sewage treatment plants show a consistent reduction in ammonia between 1970 and 1992 (Mueller and Helsel, 1996). The proportion of the US population served by wastewater treatment plants increased from 42% in 1970 to 74% in 1998 (USEPA, 1998c). A widely scattered surface-water monitoring network has shown national reductions in faecal bacterial and phosphorus concentrations (Lettenmaier *et al.*, 1991; Knopman and Smith, 1993; Smith *et al.*, 1993; Mueller and Helsel, 1996). Case studies, opinion surveys and anecdotal information suggest that these reductions in pollutants have improved the health of aquatic ecosystems in many basins, particularly near urban areas (Knopman and Smith, 1993). However, challenges to water quality remain, including continuing discharges of pollu-tants from a growing population and economy and pollution from agriculture and other non-point sources. A growing share of remaining water quality problems are due to pollution from non-point sources (USEPA, 1998c).

The most recent US Environmental Protection Agency (USEPA) National Water Quality Inventory reports indicate the nature of water quality impair-ments (Table 5.1) (USEPA, 2000a). The Water Quality Inventory is prepared with information contained in semi-annual reports from the states, required by the Clean Water Act, on the status of their surface-water resources (known as section 305(b) reports). In 1998, it was found that 35% of river miles, 45% of lake acres (excluding the Great Lakes) and 44% of estuary square miles did not fully support the uses for which they were designated by states under the Clean Water Act. States reported that agriculture is the leading source of impairment in the nation's rivers and lakes and a major source of impairment in estuaries.

Another source of information is the 1998 section 303(d) list of impaired waters, submitted to USEPA by states, tribes and territories. These are waters that do not support water quality standards and cannot meet those standards through point source controls alone. For the 1998 listing cycle, 21,845 waters with 41,318 associated impairments were listed, covering over 300,000 miles (483,000 km) of rivers and streams and more than 5 million acres (2 million ha) of lakes (USEPA, 2000b). Over 218 million people live within 10 miles of a polluted waterbody. The top three categories of impair-ment identified in the section 303(d) lists are sediments (6133 waters), pathogens (5281 waters) and nutrients (4773 waters). Pesticides ranked tenth, affecting 1432 waters. Reductions in non-point source pollution loads will be required for these waters to achieve water quality standards.

Groundwater quality is also a concern in some areas, though much less is known about the problem. Many states report on the general quality of their groundwater resources in their section 305(b) reports, although this is optional. Of 38 states that reported overall groundwater quality in 1992, 29 judged their groundwater quality to be good or excellent (USEPA, 1994c). Generally, states reported that contamination of groundwater was localized. In 1994, 45 states reported that pesticide and fertilizer applications were sources of groundwater contamination (USEPA, 1995). However, out of 37 states reporting sources of contamination in 1996, only 18 reported pesticides as a source and only 17 reported fertilizer as a source (USEPA, 1998). This difference raises questions about the usefulness of section 305(b) reports for information on groundwater quality. An important difference between groundwater and surface water is

Table 5.1. Status of US surface water quality, 1990–1996.

	Rivers (miles)				Lakes[a] (acres)				Estuaries (square miles)			
	1992	1994	1996	1998	1992	1994	1996	1998	1992	1994	1996	1998
Water systems assessed	18	17	19	23	46	42	40	42	74	78	72	32
Meeting designated uses[b]												
Supporting	62	64	64	65	56	63	61	55	68	63	62	56
Partially supporting	25	22	36[c]	35	35	28	39[c]	45	23	27	38[c]	44
Not supporting	13	14			9	9			9	9		
Clean Water Act goals: fishable												
Meeting	66	69	68	69	69	69	69	71	78	70	69	65
Not meeting	34	31	31	30	31	31	31	29	22	30	30	34
Not attainable	<1	<1	<1	<1	<1	<1	<1	<1	0	0	0	0
Clean Water Act goals: swimmable												
Meeting	71	77	79	72	77	81	75	80	83	85	84	91
Not meeting	20	23	20	24	22	19	25	29	17	15	16	9
Not attainable	9	<1	<1	5	<1	<1	<1	<1	0	<1	<1	0

[a] Excluding Great Lakes.
[b] Supporting: water quality meets designated use criteria; partially supporting: water quality fails to meet designated use criteria at times; not supporting: water quality frequently fails to meet designated use criteria.
[c] In 1996, the categories 'Partially supporting' and 'Not supporting' were combined.
Source: USDA ERS, based on USEPA National Water Quality Inventories (1994c, 1995, 1998b, 2000a).

that, once polluted, groundwater remains contaminated for a much longer period of time. This makes pollution of groundwater a more serious problem when the resource is utilized for economic purposes, such as drinking water.

The most comprehensive assessment of groundwater quality is USEPA's National Survey of Pesticides in Drinking Water Wells, conducted over 1988–1990. This survey tested for some common groundwater pollutants besides pesticides. USEPA found nitrate in more than half of the 94,600 community water system (CWS) wells and almost 60% of the 10.5 million rural domestic drinking-water wells, making nitrate the most frequently detected chemical in well water (USEPA, 1992a). The survey also found that 10% of the CWS wells and 4% of rural domestic wells contained at least one pesticide.

An evaluation of data from over 300 studies of pesticide occurrence in groundwater found that pesticides or their transformation products had been detected in groundwaters of more than 43 states (Barbash and Resek, 1996). At least 143 different pesticides had been detected.

Costs of Water Quality Impairments

Impairments to water quality from agricultural and other pollutants reduce the ability of water resources to provide services to water users. Some estimates of the magnitude of these costs exist, but there is currently a poor understanding of the costs of water pollution (see Chapter 4 for further discussion).

Annual costs to the water treatment industry from sediment were estimated to be between $458 million and $661 million in 1984 (Holmes, 1988). Reservoir sedimentation is one of the consequences of soil erosion. Survey data collected by US Department of Agriculture (USDA) and US Department of Interior (USDI) indicated that in the 1970s and early 1980s sedimentation eliminated slightly more than 0.2% of the nation's reservoir capacity each year (Crowder, 1987). Annual economic costs, based on replacing lost capacity, were estimated to be $819 million per year (Crowder, 1987). Sedimentation in navigation channels increases the costs to shipping by increasing transit time and decreasing the amount of cargo that can be carried. The Army Corps of Engineers incurred dredging costs of over $500 million per year for maintaining navigation channels over the period 1992–1998 (C. Davison, 2000, personal communication). Sediment damages from agricultural erosion have been estimated to be between $2 billion and $8 billion per year (Ribaudo, 1989). These estimates include damages or costs to navigation, reservoirs, recreational fishing, water treatment, water conveyance systems, and industrial and municipal water use.

There are very few estimates of the damages caused by nutrients in water resources, despite their pervasive impacts. Since the impacts on quality are felt through complex biological relationships, it has been difficult to determine the effects of nutrients. There are some economic estimates related to drinking water that give an indication of the damages from nitrate contamination of drinking water sources. USEPA (1997a) estimated that a total investment of $200 million is needed for additional drinking water treatment facilities to meet federal nitrate standards. Crutchfield *et al.* (1997) estimated total consumer willingness-to-pay for reduced nitrate in drinking water in four US watersheds (White River, Indiana; Central Nebraska; Lower Susquehanna; Mid-Columbia Basin) to be about $314 million per year. The benefits of nitrate-free drinking water were estimated to be $351 million.

Pesticide residues reaching surface-water systems may harm freshwater and marine organisms, damaging recreational and commercial fisheries (Pait *et al.*, 1992). Pimentel *et al.* (1991) estimated that direct annual losses from fish kills due to pesticides were less than $1 million, though the authors considered their result an underestimate. The cost to water suppliers of providing drinking water that meets Safe Drinking Water Act standards can increase substantially when pesticides are present in the water source. For example, the cost to 11 small water suppliers in the Midwest of installing additional water treatment to remove the herbicide atrazine from drinking water was estimated to be $8.3 million in capital costs, and $180,000 per year in operating costs (Langemeier, 1992). USEPA (1997a) estimates that total costs for additional treatment facilities needed to meet current regulations for pesticides and other specific chemicals would be about $400 million, with about another $100 million required over the next 20 years.

Total damages from salinity in the Colorado River range from $310 million to $831 million annually, based on the 1976–1985 average levels of river salinity. These include damages to agriculture ($113–122 million), households ($156–638 million), utilities ($32 million) and industry ($6–15 million) (Lohman *et al.*, 1988).

Outbreaks of waterborne diseases are a growing concern. USEPA (1997a) estimated the cost of facilities for improved microbial treatment to be about $20 billion over the next 20 years, with about half of that needed immediately. The health cost of *Giardia* alone is estimated to be between $1.2 billion and $1.5 billion per year (USEPA, 1997b). *Cryptosporidium* is a more recently identified threat, with oocysts present in 65–97% of surface water sampled in the United States (CDC, 1996). The organism has been implicated in gastroenteritis outbreaks in Milwaukee, Wisconsin (400,000 cases and 100 deaths in 1993), and in Carrollton, Georgia (13,000 cases in 1987). The cost of the Milwaukee outbreak is estimated to have exceeded $54 million (Anonymous, 1994). While the source of the organism in these outbreaks was never determined, its occurrence in livestock herds has brought some attention to this sector, especially given the proximity of cattle and slaughterhouses to Milwaukee (MacKenzie *et al.*, 1994).

Current US Framework for Water Quality Protection

The *Clean Water Act* (CWA) is the major federal statute that addresses water quality (Davies and Mazurek, 1998). When passed in 1972 as the Federal Water Pollution Control Act (PL 92-500) it established the current structure of federal, and to a large extent, state water quality protection. It has been amended three times, the latest in 1987. Its sections establish programmes for reducing pollution from both point sources and non-point sources. While the Act is aimed primarily at protecting surface water, it also provides for the development of federal, state and local programmes for reducing and preventing contamination of groundwater (USEPA, 1998b).

Section 402 of the CWA established the Point Source Program, which is aimed at restricting the discharge of pollutants from municipal and industrial dischargers. The basis of the programme is the National Pollutant Discharge Elimination System (NPDES) permit. Each point-source discharger must obtain a discharge permit before it can discharge into surface water. The permit requires point-source dischargers to comply with technology-based controls (uniform USEPA-established standards of treatment that apply to certain industries and municipal sewage treatment facilities) or water quality-based controls that invoke state numeric or narrative water quality standards (Moreau, 1994). Large, confined animal operations (over 1000 animal units) fall under the NPDES. Currently, 43 states manage their own NPDES programmes. Over 500,000 discharge sources are subject to NPDES permits (USEPA, 1998b).

When technology-based controls are inadequate for water to meet state water quality standards, section 303(d) requires states to identify those waters and to develop total maximum daily loads (TMDL) (USEPA, 1993). Federal regulations and USEPA guidance for TMDL implementation describe a process where regulators establish wasteload allocations (WLA) for point sources and load allocations (LA) for nonpoint sources and natural sources (Bartfeld, 1993). Together, WLAs and LAs comprise the TMDL, or the maximum discharge of pollutant in the basin that will allow the water quality standard to be met (Graham, 1997). A necessary component of this process is the identification of all loads and an assessment of the assimilative capacities of the waterbody, in relation to the water quality standards to be met. USEPA has responsibility for developing TMDLs if a state fails to act (USEPA, 1993). Over 500 TMDL plans have been initiated since 1992, and 225 have been completed and approved by USEPA (1997d).

Section 319 of the Clean Water Act established the non-point source control programme for addressing polluted runoff from land surfaces. It consists of a three-stage national programme that is implemented by the states with federal approval and assistance (USEPA, 1998b). States address non-point source pollution by: (i) assessing state waters to determine where non-point source pollution is impairing uses; (ii) developing a non-point source management programme; and (iii) implementing the management programme. All states currently have USEPA-approved management programmes. The states are free to choose the policy instruments contained in the management plans. Most states are relying on voluntary approaches that emphasize education, technical assistance and economic incentives.

A second federal statute that directly addresses non-point source pollution is the *Coastal Zone Act Reauthorization Amendments* (CZARA), which established a coastal non-point source pollution control programme. The National Oceanic and Atmospheric Administration (NOAA) of the Department of Commerce administers the programme. Section 6217 requires that states with federally approved coastal zone management programmes develop and implement coastal non-point pollution control programmes to ensure protection and restoration of coastal waters (USEPA, 1998b). Thirty states are required to develop such plans and all have submitted non-point source programmes for USEPA and NOAA approval.

Under CZARA, state coastal non-point source programmes must provide for the implementation of best management practices specified by USEPA in national technical guidance. A list of economically achievable measures for controlling agricultural non-point source pollution is part of each state's management plan. In addition, more stringent management measures must be developed by each state as necessary to attain and maintain water quality standards where the national technical guidance is inadequate for meeting water quality goals. Federal guidance is broad enough to allow states to identify management measures that are best suited for local conditions, thus avoiding the inefficiencies of requiring practices with 'national' standards. However, state management plans must still be approved by USEPA and NOAA, providing some measure of quality assurance over state management measures. States can first try voluntary incentive mechanisms to promote adoption, but must be able to enforce management measures if voluntary approaches fail. Implementation of CZARA plans is not required until 2004, and the penalties for failing to implement plans are small (loss of some funding). Coastal zone management plans must be incorporated into states' section 319 non-point source programmes.

The Clean Water Act and CZARA focus primarily on surface water. Groundwater is addressed by four federal statutes. The CWA encourages groundwater protection (section 102), recognizing that groundwater provides a significant portion of based flow to streams and lakes. Specifically, the CWA provides for the development of federal, state and local comprehensive programmes for reducing, eliminating and preventing groundwater contamination. The CWA provides a framework for states to develop their own programmes, rather than specifying or requiring specific actions to be taken.

The *Safe Drinking Water Act* (SDWA) requires the USEPA to set standards for drinking water quality and requirements for water treatment by public water systems (Morandi, 1989). The SDWA authorized the *Wellhead Protection Program* (WHP) in 1986 to protect supplies of groundwater used as public drinking water from contamination by chemicals and other hazards, including pesticides, nutrients and other agricultural chemicals (USEPA, 1993). The programme is based on the concept that land-use controls and other preventive measures can protect groundwater. As of December 1998, 45 states had a USEPA-approved wellhead protection programme (USEPA, 1998e). The 1996 amendments to the SDWA require USEPA to establish a list of contaminants for consideration in future regulation (USEPA, 1998a). The *Drinking Water Contaminant Candidate List*, released in March 1998, lists chemicals by priority for: (i) regulatory determination, (ii) research and (iii) occurrence determination. Several agricultural chemicals, including meto-lachlor, metribuzin and the triazines, are among those to be considered for potential regulatory action (USEPA, 1998a). USEPA will select five contami-nants from the 'regulatory determination priorities' list and determine by August 2001 whether to regulate them to protect drinking-water supplies.

Also under the 1996 amendments, water suppliers are required to inform their customers about the level of certain contaminants and associated USEPA standards, and the likely source(s) of the contaminants, among other items (USEPA, 1997c). If the supplier lacks specific information on the likely source(s), set language must be used for the contaminants, such as 'runoff from herbicide used on row crops' (e.g. for atrazine). 'The information con-tained in the consumer confidence reports can raise consumers' awareness of where their water comes from, ... and educate them about the importance of preventative measures, such as source water protection' (*Federal Register*, 19 August 1998, p. 44512). Increased consumer awareness concerning water supplies could lead to public pressure on farmers to reduce pesticide use (Smith and Ribaudo, 1998).

The *Source Water Assessment Program* (SWAP) was enacted in the 1996 SDWA amendment and builds upon the WHP. Under SWAP, states must delin-eate the boundaries of areas providing source waters for public water systems, identify the origins of regulated and certain unregulated contami-nants in the delineated areas, and determine the susceptibility of public water systems to such contaminants.

The *Sole Source Aquifer Protection Program* allows communities, individuals and organizations to petition USEPA to designate aquifers as the 'sole or principal' source of drinking water for an area. Once an aquifer receives this designation, USEPA has the authority to review and approve projects receiving federal financial assistance that may contaminate the aquifer. This includes projects supported by the US Department of Agriculture.

The *Underground Injection Control Program* established federal regulation of underground injection wells, which are used to dispose of hazardous or non-hazardous fluids by pumping them into underground rock formations. Regulations are for ensuring that the disposed fluids do not pose a contamina-tion risk to drinking water sources. This programme pertains mainly to indus-trial discharges.

Another principal programme for controlling sources of pollution that may contaminate groundwater is the *Resource Conservation and Recovery Act* (RCRA). This law covers underground storage tanks and solid waste and the treatment, storage and disposal of hazardous wastes. The intent of the Act is to protect human health and the environment by establishing a comprehensive regulatory framework for investigating and addressing past, present and future environmental contamination. This is done through a process of identifying wastes that pose hazards if improperly managed, and establishing requirements for waste treatment and management to ultimate disposal.

The *Comprehensive Environmental Response, Compensation, and Liability Act* (CERCLA) created several programmes operated by USEPA and the states for protecting and restoring contaminated groundwater, including the Superfund programme. The goal is to return usable groundwater to beneficial uses, wherever possible, within a reasonable time frame (USEPA, 1998b). The law is designed primarily for addressing point sources of ground-water contamination, such as landfills.

Another law that is based on a class of potentially hazardous products is the *Federal Insecticide, Fungicide and Rodenticide Act* (FIFRA) (PL 92-516). Under the law, USEPA is responsible for registering pesticides for specified uses on the basis of both safety and benefits before they may be sold, held for sale, or distributed in commerce (USGAO, 1991a). USEPA can register a pesticide only if it determines that the pesticide will perform its intended function with-out causing any unreasonable risk to human health or the environment. Economic, social and environmental benefits and costs of use are to be consid-ered in the registration process. Under the law, it is unlawful to use any pesticide in a manner inconsistent with its labelling, which must clearly specify directions for use, including application and concentration rates, the pests and/or plants on which the pesticide is intended for use, when it can be applied, and timing between applications. Water quality has been a major consideration in the regis-tration of some pesticides, such as atrazine (Ribaudo and Bouzaher, 1994).

USDA implements a variety of programmes related to water quality that directly involve agricultural producers. These programmes use financial, educational, and research and development tools to help farmers voluntarily adopt management practices that protect water quality and achieve other environmental objectives. The *Environmental Quality Incentives Program* (EQIP), authorized by the Federal Agriculture Improvement and Reform Act of 1996, provides technical, educational and financial assistance to eligible farmers and ranchers to address soil, water and related natural resource concerns on their lands in an environmentally beneficial and cost-effective manner (USDA ERS, 1997). This programme consolidated the functions of a number of USDA programmes, including the Agricultural Conservation Program, Water Quality Incentives Program, Great Plains Conservation Program and Colorado River Basin Salinity Program. The objective of EQIP is to encourage farmers and ranchers to adopt practices that reduce environ-mental and resource problems. Assistance is targeted to priority conservation areas and to specified problems outside of those areas. Five- to 10-year contracts with landowners may include incentive payments as well as cost-sharing of up to 75% of the costs of installing approved practices. Fifty per cent of the funding available for the programme is to be targeted at natural resource concerns related to livestock production. However, owners of large, concentrated livestock operations are not eligible for cost-share assistance for installing animal waste storage or treatment facilities. There is general statutory guidance to manage EQIP so as to maximize environmental benefits per dollar expended.

The *Water Quality Program* (WQP), established in 1990 and currently winding down, attempted to determine the precise nature of the relationship between agricultural activities and water quality. It also attempted to develop and induce adoption of technically and economically effective agrochemical management and production strategies that protect surface- and ground-water quality (USDA, 1993). WQP included three main components: (i) research and development; (ii) educational, technical and financial assistance; and (iii) database development and evaluation. The first two components were carried out in targeted projects designed to address a particular water quality problem. Seven projects were devoted to research and development (Management System Evaluation Areas) and 242 projects were devoted to assisting farmers to implement water quality-enhancing farming practices (Hydrologic Unit Area projects, Water Quality Incentive projects, Water Quality Special projects and Demonstration Projects). Despite the large expenditures on conservation practices, little documentation of impacts on water quality was made and no estimates of economic benefits from water quality improvements were made.

Conservation Compliance provisions were enacted in the Food Security Act of 1985 to reduce soil erosion (USDA ERS, 1997). Producers who farm highly erodible land (HEL) were required to implement a soil conservation plan to remain eligible for other specified USDA programmes that provide financial payments to producers. Violation of the plan would result in the loss of price support, loan rate, disaster relief, Conservation Reserve Program (CRP) and Farmers' Home Administration (FmHA) benefits. While this is not intended to be a pollution prevention programme, reducing soil erosion has implications for water quality.

Conservation compliance, described in Chapter 3, has reduced soil erosion that might have been impairing water quality. Annual soil losses on HEL cropland have been reduced by nearly 900 million tons (USDA NRCS, 1996). If conservation plans were fully applied on all HEL acreage, the average annual soil erosion rate would drop from 16.8 to 5.8 tons per acre per year (USDA NRCS, 1996).

Conservation compliance has been calculated to result in a large social dividend, primarily due to offsite benefits. An evaluation using 1994 HEL data indicates that the national benefit/cost ratio for compliance is greater than 2:1 (although the ratios vary widely across regions) (USDA ERS, 1994). In other words, the monetary benefits associated with air and water quality and productivity outweigh the costs to government and producers by at least 2:1 (USDA ERS, 1994). Average annual water quality benefits from conservation compliance were estimated to be about $13.80 per acre (USDA ERS, 1994). However, these findings do not necessarily indicate that existing compliance programmes are cost-effective non-point pollution control mechanisms.

Water quality would also be expected to improve from two USDA land-retirement programmes. The *Conservation Reserve Program* was established in 1985 as a voluntary long-term cropland retirement programme (USDA ERS, 1997). The CRP uses subsidies to retire cropland from production where it is believed to be especially prone to producing environmental problems. In exchange for retiring highly erodible or other environmentally sensitive cropland for 10–15 years, CRP participants are provided with an annual per acre rent and half the cost of establishing a permanent land cover (usually grass or trees). Payments are provided for as long as the land is kept out of production.

CRP eligibility has been based on soil erosion (first nine sign-ups) and potential environmental benefits (sign-ups ten and up) to retire cropland that has been generating the greatest environmental damages. Starting in the tenth sign-up, the cost-effectiveness of CRP outlays in achieving environmental goals was increased by using an environmental benefits index (EBI) to target funds to more environmentally sensitive areas. The EBI measures the potential contribution to conservation and environmental programme goals that enrolment bids would provide. The seven coequal conservation and environmental goals targeted include: surface water quality improvement; groundwater quality improvement; preservation of soil productivity; assistance to farmers most affected by conservation compliance; encouragement of tree planting; enrolment in established USDA Water Quality Program projects; and enrolment in established conservation priority areas. Enrolment bids with a higher ratio of EBI to rental payment were accepted ahead of bids with lower ratios. Thus, at least to some degree, the use of EBI ensures that land with characteristics most related to environmental quality are enrolled first.

The CRP has converted a total of 36.4 million acres of cropland (14.7 million ha – about 8% of all US cropland) to conservation uses since 1985. Net social benefits of the CRP are estimated to range between $4.2 billion and $9 billion (Hrubovcak *et al.*, 1995).

The *Wetlands Reserve Program*, authorized as part of the Food, Agriculture, Conservation, and Trade Act of 1990, is primarily a habitat protection programme, but retiring cropland and converting back to wetlands also has water quality benefits (USDA ERS, 1997). These benefits include not only reduced chemical use and erosion on former cropland, but also the ability of the wetland to filter sediment and agricultural chemicals from runoff and to stabilize stream banks. The water quality benefits from WRP have not been estimated.

In addition to the above programmes that provide direct assistance to producers, USDA provides assistance to state agencies and local governments through the *Small Watershed Program* (otherwise known as Public Law 566) (USDA ERS, 1994). To help to prevent floods, protect watersheds and manage water resources, this programme includes establishment of measures to reduce erosion, sedimentation and runoff.

Evolution of Non-point Source Policy

The history of non-point source policy in the US gives some indication of the difficulties inherent in its control. A centralized control policy would have to account for many different combinations of factors to be efficient. Non-point sources of water pollution were first identified as necessary for control in the 1972 amendments to the Federal Water Pollution and Control Act, known popularly as the Clean Water Act (CWA). Although the emphasis of the CWA was on control of pollution from point sources, such as factories and sewage treatment plants, section 208 called for the development and implementation of 'areawide' water quality management programmes to ensure adequate control of all sources of pollutants, point and non-point, in areas where water quality was impaired. The CWA directed states to develop plans for reducing non-point source (NPS) pollution, including appropriate land management controls. The 1977 amendments further emphasized the role of NPS control in meeting water quality goals, but did not change the basic approach.

The section 208 process is generally not seen as being a success (Harrington *et al.*, 1985; USEPA, 1988; Cook *et al.*, 1991; USGAO, 1991b). A series hearing by the US Congress, House of Representatives, Public Works and Transportation Subcommittee found that technical and financial support for the programme was lacking, coordination with the point source programme was non-existent and data necessary for implementing an effective programme were inadequate (Copeland and Zinn, 1986). The consequence was that states lagged in the development of areawide management programmes and USEPA could not readily judge whether the section 208 plans finally developed were adequate for achieving NPS goals. USEPA was also not given effective enforcement tools to ensure that NPS management plans were truly viable, or actually implemented (Wicker, 1979).

Part of the reason for the lack of progress was the perception that non-point source pollution was not particularly important. Point source pollution was seen as the more serious problem, being responsible for the most visible water quality problems. Point sources were also easier to control through centralized technology standards, in the form of the NPDES permit system. As a result, greater effort and resources were devoted to point source pollution, with little dissent from environmental or other groups.

By the mid-1980s, USEPA started to take a harder look at non-point source pollution as an important cause of remaining water quality problems. While point source discharges were still causing problems, non-point source pollution had become the largest unregulated source of pollution. In its Report to Congress, USEPA (1984) stated: 'In many parts of the country, pollutant loads from non-point sources present continuing problems for achieving water quality goals and maintaining designated uses' (p. 1-1). The report also singled out agriculture as 'the most pervasive cause of non-point source water quality problems' (p. 2-6).

Congress responded by revamping the non-point source programme in the Water Quality Act (WQA) of 1987. The WQA placed special emphasis on non-point source pollution by amending the Clean Water Act's Declaration of Goals and Policy to focus on the control of non-point sources of pollution (USEPA, 1988). The WQA also added section 319, which is described above. Even though the goals of non-point source pollution control policy were clarified and a more structured framework was introduced, states are still responsible for developing non-point source programmes, including the specific instruments to be used. This is in direct contrast to the way in which point sources are addressed.

Passing responsibility for non-point source pollution to states has both advantages and disadvantages with respect to the efficiency of control. On the positive side, states are closer to the problem, would be in a better position to collect the information necessary to implement efficient programmes and would be more likely to fashion a response appropriate to the problem. A basic principle of the economic theory of federalism is that economic efficiency in the provision of public goods is generally best served by delegating responsibility for the provision of the good to the lowest level of government that encompasses most of the associated benefits and costs (Shortle, 1995). In many cases the impacts of non-point source pollution are most pronounced close to its point of origin.

The characteristics of non-point source pollution vary over geographical space, due to the great variety of farming practices, land forms, climate and hydrological characteristics found across even relatively small geographical areas. A centralized control policy would have to account for many different combinations of factors. Efficiency can generally be obtained only through exceedingly high information and administration costs. Reducing these costs through national standards comes at a price of reduced efficiency. Decentralized policies need to account for less variation and would presumably entail lower administration costs.

On the negative side, states may lack the fiscal resources to implement the monitoring necessary to develop an efficient non-point source pollution policy. States may also lack the technical expertise for devising efficient non-point source pollution control policies. They are also in a poor position to handle water pollution from upstream dischargers in other states. While many of the problems from non-point source pollution are felt close to the source, some non-point source pollutants can travel long distances in major rivers or affect regional waterbodies such as the Gulf of Mexico or Chesapeake Bay. The beneficiaries of a state's pollution control policies could therefore be residents of other states. There are very few examples where states have come together without federal prodding to address regional water quality issues, despite common goals and the fact that an individual state may not be able to meet its water quality goals without better control of interstate pollution.

A consequence of the different approaches taken for point and non-point source pollution is that gains in water quality have come at a higher cost than if both sources had been treated more evenly. In an assessment of the benefits and costs of the Clean Water Act, Freeman (1994) concluded that costs are very likely substantially in excess of realized benefits. One of the reasons was the focus on point sources, even when non-point sources were significant contributors to water quality impairments. Additional evidence of inefficiencies in the Clean Water Act comes from comparisons of marginal pollutant removal costs between point and non-point sources. Nationally, allowing point sources at 470 sites to reduce treatment costs by enabling them to purchase nutrient reductions from non-point sources in a point–non-point trading programme would save dischargers between $611 million and $5.6 billion (USEPA, 1994a). This means that a water quality goal at these sites can be achieved at a lower cost by reducing non-point source discharges rather than point source discharges.

State Programmes

All states provide incentives to farmers for adopting management practices that reduce agricultural non-point source pollution. Common strategies include education programmes, technical assistance programmes and cost-sharing for implementation of prevention and control measures. These approaches are generally modelled after USDA's programmes.

Recently, more states have been moving beyond a voluntary approach to address non-point source pollution towards mechanisms designed to enforce certain behaviour. These 'enforceable mechanisms' include regulation and liability provisions (ELI, 1997).

State laws using enforceable mechanisms for non-point source pollution vary widely in definitions, enforcement mechanisms, scope and procedures. States are taking very dissimilar directions in enforceable non-point source pollution control policy, largely because of the absence of federal direction (ELI, 1997). Some of the catalysts moving states towards stronger measures include: immediate problems that have demanded attention (e.g. nitrate contamination of groundwater in Nebraska; animal waste problems in North Carolina; pesticide contamination of groundwater in California and Wisconsin); the use of TMDLs for identifying sources of water contaminants; the requirements of the CZARA; and the improving technical ability of states to assess their waters (ELI, 1997).

State laws using enforceable mechanisms for non-point source pollution vary widely in definitions, enforcement mechanisms, scope and procedures (ELI, 1997). The mechanisms that states are using to make adoption of best management practices (BMPs) more enforceable can be grouped into five categories (ELI, 1997): (i) making BMPs directly enforceable in connection with required plans and permits; (ii) making BMPs enforceable if the operator is designated a 'bad actor'; (iii) making compliance with BMPs a defence to a regulatory violation; (iv) making BMPs the basis for an exemption from a regulatory programme; and (v) making compliance with BMPs a defence to nuisance or liability actions. All these mechanisms focus on BMPs (design-based) rather than water quality measures (performance-based) and this is a direct result of non-point source pollution discharges being unobservable. Policies are focusing on those factors in the pollution process that can be observed. Some states are trying to link BMPs more closely with observed problems by employing 'triggers' (measured environmental conditions). The types of BMPs required depend on the level of pollutants found in water through monitoring. In a few cases, where the relationship between production activities and water quality are better understood and can be monitored, policies have adopted management measures that are more performance-based.

Table 5.2 summarizes some of the foci of such mechanisms being used by states. While many states have provisions that deal with water quality as it relates to agricultural non-point source pollution, they often target only a subset of water quality problems. Few states deal with agricultural non-point source pollution in a comprehensive manner. Most target individual pollutants (e.g. sediment), resources (e.g. groundwater), regions (e.g. coastal zone), or type of operations (e.g. swine). Many of these laws have been enacted within the past 5 years and so the impacts of these policies on producers or on the environment have yet to be seen. The following are some examples of the approaches being taken by states.

Groundwater protection from pesticides – California

Pesticides in groundwater are a major concern in California. Intensive production of fruit, vegetables and other crops requires the application of a wide variety of pesticides. California's groundwater is a major resource vital to the economic development of the state (Holden, 1986). California is using the Pesticide Contamination Prevention Act (Division 7, Chapter 2, Article 15, FAC) to protect groundwater from pesticide pollution. The state has created a groundwater protection list of pesticides subject to regulation. Inclusion on the list is determined by the physical characteristics of the chemical. The law requires the state to set numeric values for six chemical characteristics that define the chemical's ability to leach into groundwater. During the registration process the manufacturer must submit information on these characteristics. If the value for one of the characteristics exceeds the prescribed numeric value, the chemical is placed on the groundwater protection list.

Pesticides on the list are regularly monitored in the environment. If the pesticide or potentially toxic degradates are found either in groundwater, or 8 ft (2.4 m) below the surface, or below the root zone, or below the zone of microbial activity, the pesticide is subject to restrictions in use or to cancellation. If it is determined that legal use of the chemical does not threaten to pollute groundwater anywhere in the state where it may be used, then use can continue without change. If current legal use is determined to pose a threat, use is allowed to continue if the label can be modified so that there is a high probability that groundwater contamination will not occur. This includes the establishment of pesticide management zones where the use of certain chemicals is restricted. If use cannot continue under alternative practices, then the chemical can be banned. However, if it is determined that cancellation or modified use will cause a severe economic hardship to agriculture, and no substitute products or practices can be effectively used, then use can continue subject to meeting water quality standards that are believed to represent acceptable risks. If continued use is allowed under the above restrictions, and groundwater contamination is found after 2 years, the chemical will be cancelled if it is carcinogenic, mutagenic, teratogenic or neurogenic.

California's law is designed to minimize statewide economic hardships. However, it is possible that farmers within pesticide management zones may face increased production costs and/or greater risk of pest losses, placing them at a competitive disadvantage to neighbouring producers. Also, it is

Table 5.2. Summary of foci of state enforceable mechanisms for controlling agricultural pollution.[a]

State	Fertilizer	Pesticides	Sediment	CAFOs/animal waste	Comprehensive
Arizona	x	x		x	
Arkansas		x		x	
California		x			
Colorado	x			x	
Connecticut	x			x	
Florida	x	x		x	
Idaho	x	x			
Illinois				x	
Indiana			x		
Iowa	x	x		x	
Kansas		x		x	
Kentucky					x
Maine					x
Maryland	x		x		x
Michigan	x				
Minnesota				x	
Montana	x	x			
Nebraska	x		x		
New York			x		
North Carolina				x	
Ohio				x	
Oregon					x
Pennsylvania			x	x	
South Dakota				x	
Virginia				x	
Washington				x	
West Virginia				x	
Wisconsin	x	x		x	
Wyoming				x	

[a] Mechanisms may apply only under certain conditions or in certain localities.

possible for important chemicals to be banned if they continue to be found after 2 years and they pose health risks to humans.

Seven herbicides had verified detections in California groundwater in 1997: atrazine, bromacil, diuron, hexozinone, norflurazon, prometron and simazine. To protect groundwater from these pesticides, 92 pesticide management zones have been established in ten counties. Growers operating within these zones are denied access to the banned pesticides through the registration process. The state has also developed a geographic information system (GIS) that enables the permit issuer to determine whether a grower's field is within a pesticide management zone.

Groundwater protection from atrazine – Wisconsin

Another example is Wisconsin's programmes for protecting groundwater from pesticides. The legal basis for these programmes is the Wisconsin Groundwater Law (1983) (Wisc. Stats., Chapter 160). The Groundwater Law requires the state to undertake remedial and preventive actions when concentration 'triggers' are reached in groundwater for substances of public health concern, including a number of pesticides. Two triggers are established for each chemical: an enforcement standard and prevention action limit (PAL). The PAL is 10, 20 or 50% of the enforcement standard, depending on the toxicological characteristics of the substances. When a PAL is exceeded, a plan for preventing further degradation is prepared. When the enforcement standard is exceeded, the chemical is prohibited in that area overlaying the aquifer that is contaminated.

An example of how the law is implemented involves the herbicide atrazine. The enforcement standard for atrazine is 3.0 ppb, and the PAL is 0.35 ppb. Well monitoring found that atrazine concentrations in many areas of the state where it is used were above the PAL (Wolf and Nowak, 1996). In some areas concentrations were above the enforcement standard. Part of the plan for addressing the problem was the passage of the Atrazine Rule, which established maximum atrazine application rates and conditional use restrictions for the state (Wisc. Admin. Code, Agri. Trade & Cons. Prot. Ag30). The Rule also established a series of zones where additional restrictions are imposed on top of the statewide rules. The result is a three-tiered management plan: statewide atrazine restriction; Atrazine Management Areas (AMA) where concentrations exceed the PAL; and Atrazine Prohibition Areas where concentrations are above the enforcement level.

Statewide atrazine restrictions consist of soil-based maximum application rates, restrictions on when atrazine can be applied and a prohibition on applying through irrigation systems. Further restrictions are placed on application rates in the AMAs. In 1993, six atrazine management areas had been established, in response to detections of atrazine in groundwater at concentrations at or greater than 0.35 ppb (Wolf and Nowak, 1996). In addition, 14 atrazine prohibition areas had been established, in response to concentrations in groundwater greater than 3.0 ppb.

An assessment of the Atrazine Rule reported that producers in AMAs were not facing an agronomic disadvantage to counterparts in non-AMAs, as represented by comparisons of yield loss predictions and assessment of weed intensity (Wolf and Nowak, 1996). However, an assessment of compliance costs was not made.

Groundwater protection from nitrogen – Nebraska

Nitrate contamination of groundwater is a serious concern in Nebraska. For years, studies have shown nitrate concentrations at varying levels in groundwater throughout the state, often much higher than the USEPA standard of 10 mg l^{-1}. Groundwater sources meet nearly all the needs of Nebraska's rural residents and 84% of the state's public water systems (Schneider, 1990).

Nebraska is divided into Natural Resources Districts (NRDs), which are local units of government charged with the responsibility of conservation, wise development and proper utilization of natural resources (Bishop, 1994). In 1982 the Nebraska legislature passed the Ground Water Management and Protection Act which allowed NRDs to establish groundwater control areas to address groundwater quality concerns (Neb. Rev. Stat. sections 46-673.01– 46-674.20). In 1986 the legislature gave NRDs the ability to require best management practices and education programmes to protect water quality. Best management practices are defined as those practices that prevent or reduce present and future contamination of groundwater, and include irrigation scheduling, proper timing of fertilizer and pesticide application, and other fertilizer and pesticide management programmes.

The Central Platte NRD used this authority to develop an aggressive groundwater protection programme for addressing a serious and growing nitrate problem in its area. Under the Central Platte regulations, areas within the management area are divided into three phases, based on current groundwater nitrate levels. A Phase I area is defined as having an average groundwater nitrate level of between 0 and 12.5 ppm; Phase II areas average between 12.6 and 20 ppm; and Phase III areas have concentrations averaging 20.1 ppm or greater.

Agricultural practices are restricted according to the level of contamination. In a Phase I area, commercial fertilizer cannot be applied on sandy soils until after 1 March. Autumn and winter applications are prohibited.

Phase II regulations include the Phase I restrictions, plus the condition that commercial fertilizer is only permitted on heavy soils after 1 November if an approved nitrogen inhibitor is used. In addition, all farm operators using nitrogen fertilizer must be certified, irrigation water must be tested annually for nitrate concentration and the content included in fertilizer recommendations, and annual reports on nitrate applications and crop yields must be filed with the NRD.

Phase III regulations combine the Phase II requirements with the requirements for split application (pre-plant and side-dress) and/or nitrogen inhibitors in the spring. In addition, deep soils analysis is required annually. Groundwater monitoring in the Central Platte NRD, which had the greatest problem, has shown a decrease in groundwater nitrate, indicating that the programme is working (Bishop, 1994).

Protection of surface water from phosphorus – Florida

Florida is home to the Everglades, a vast wetland containing a multitude of unique wildlife species. These wetlands have been degraded over time by human activities, including drainage, development and agriculture. Phosphorus loadings to the Everglades ecosystem upset the nutrient balance and promote the growth of undesirable plant species. A strategy has been developed for reducing phosphorus loadings to the Everglades. Much of the phosphorus is coming from the agricultural areas to the north of the Everglades.

Animal waste from dairy operations around Lake Okeechobee has been identified as a major source of phosphorus loadings. A series of three regulatory policies was applied to the Lake Okeechobee basin to reduce these loadings (Schmitz *et al.*, 1995). The Dairy Rule technology standard required the collection, storage and treatment of wastewater from dairy operations (Florida Admin. Code 62-670.500). As an alternative to complying with the Dairy Rule, operators could choose to enrol in the dairy buy-out programme, under which operators were offered a one-time payment for moving their operations out of the basin and accepting an easement on the land. The third policy, known as the Works of the District Rule, imposed a maximum allowable phosphorus concentration in runoff performance standards for dairies (Florida Admin. Code 62-670.500). Such an approach is possible because the extensive system of drainage ditches enables the monitoring of phosphorus discharges from individual sources.

The imposition of regulations resulted in direct cost, opportunity cost of the operator's time, waiting cost and regulatory uncertainty cost (Schmitz *et al.*, 1995). The implementation of the three regulatory programmes was estimated to have cost $41.4 million. About half the costs were incurred by the dairy industry, the rest by the government. The dairy buy-out programme reduced the region's cow herd by 14,000 animals. For the dairies that remained in operation, the Dairy Rule and the Works of the District Rule increased average cost of production by $1.15 cwt^{-1}. Annualized investment costs of compliance were estimated at $0.97 cwt^{-1}. Annual operation and maintenance costs were estimated to range between $0.14 and $0.2 cwt^{-1} (1 cwt = 0.05 tonnes). The targeting of the regulations to a particular geographical area resulted in a shift in milk production to other regions of the state.

A basin-wide incentive programme for the Everglades was initiated in the 1994 Everglades Forever Act to reduce phosphorus loadings from cropland – primarily vegetables and sugarcane (F.S. 373.4592). The law mandates a 25% reduction in phosphorus loads discharging from the Everglades Agricultural Area (EAA) between Lake Okeechobee and the Everglades National Park. The Act requires farmers to prepare plans and to install BMPs by the beginning of 1995. BMPs include soil testing, applying fertilizer directly to crop roots, providing for longer drainage retention, sediment controls and innovative crop location. Associated with the Act is the Agricultural Privilege Tax, which is aimed at increasing the discharge reductions beyond 25%. A tax, starting at $24.89 per acre, was put on all crop acres in the EAA. The tax will increase every 4 years to a maximum of $35.00 per acre from 2006 through 2014, unless farmers in the EAA exceed an overall 25% basin-wide phosphorus reduction goal. Revenue from the tax is earmarked for the construction of Stormwater Treatment Areas – essentially constructed wetlands for removing phosphorus before it reaches the Everglades National Park.

Protection of estuary from nutrients – Maryland

Chesapeake Bay is a vital resource of the mid-Atlantic coast. It provides habitat for many species of animals and plants, including commercially valuable fish and shellfish. Over the years nutrient enrichment and other pollutants have degraded its condition, greatly decreasing the catch of fish and shellfish. Maryland, Virginia, Pennsylvania and the District of Columbia have promised to implement programmes reducing nutrient loads to the Bay by 40% and to reduce sediment and pesticides as well. Maryland has invoked several pollution control programmes aimed at agriculture, primarily in response to concerns over the health of Chesapeake Bay. Maryland's Chesapeake Bay Critical Area Program requires all agricultural land in the Critical Area (all land within 1000 ft – 304 m – of the Bay or a tributary) to have a Soil Conservation and Water Quality (SCWQ) plan (Md. Code Ann., Nat. Res., section 8-1801 *et seq.*). An SCWQ plan is the standard conservation planning tool in Maryland and it addresses all agricultural non-point source pollution on farms. SCWQ plans call for the implementation of BMPs for sediment, nutrients and pesticides. The appropriate management practices are determined on a farm-by-farm basis and are selected from the NRCS Field Office Technical Guide. As applied to the Critical Area Program, planning is done for the entire farm, not just for the parts of fields within the Critical Area. Approximately 34% of the land in the coastal zone is covered by an SCWQ plan.

In 1998 Maryland passed its Water Quality Improvement Act. This is one of the most comprehensive farm nutrient control laws in the country. Under the law, most farming operations must have and implement a nitrogen- and phosphorus-based nutrient management plan. The plan covers all sources of nutrients, including animal waste and sewage sludge. Details of how the law will actually be implemented are still being worked out.

Protection of surface waters from agricultural pollution – Vermont

An example of a comprehensive water quality law that includes requirements for general adoption of technology standards is Vermont's Agricultural Nonpoint Source Pollution Reduction Program, which uses a three-level approach (Vermont Statutes Annotate Title 6, Chapter 215). The law requires all farmers to follow a set of 'accepted agricultural practices' (AAPs). The statewide restrictions are designed to reduce non-point pollution through implementation of improved farming practices. The law requires that these practices be technically feasible as well as cost-effective for farmers to implement without financial assistance. AAPs cover the range of agricultural pollutants that can enter surface water and groundwater, including sediment, nutrients and agricultural chemicals. Some examples of AAPs include erosion and sediment control, animal waste management, fertilizer management and pesticide management. Animal operations of greater than 950 animal units must use AAPs in order to obtain a permit. One practice that is mandatory for all fields bordering permanent waters is vegetative buffer strips. Under the law, all farmers in the state must follow AAPs as part of their normal operations. Implementation of AAPs creates a presumption of compliance with Vermont Water Quality Standards. Where the AAPs are insufficient to achieve water quality goals, voluntary installation of additional BMPs will be encouraged through financial assistance. If water quality continues to be a problem, then BMPs will be required on specific farms. BMPs typically require the installation of structures such as manure storage systems.

Failure to use practices considered to be consistent with the AAP rules results in a warning of non-compliance. Continued failure to adopt recommended practices can result in cease-and-desist orders and administrative penalties. The law is a little vague on how compliance is determined, but it appears that a finding of non-compliance can arise from an inspection that does not result from a citizen complaint.

Seeking greater efficiency through trading – North Carolina

North Carolina has adopted a basin-oriented water quality protection strategy that includes trading between different sources of pollution. This was made possible by North Carolina turning to the total maximum daily load provisions of the Clean Water Act (USEPA, 1997d). According to the Clean Water Act, if the technology-based point source programme fails to achieve water quality standards, a second tier of regulations would be implemented. These are based on the quality of the receiving waters and are known as the total maximum daily load (TMDL) provisions. Federal regulations and USEPA guidance for TMDL implementation describe a process where regulators establish wasteload allocations (WLA) for point sources and load allocations (LA) for non-point sources and natural sources (Bartfeld, 1993). Together,

WLAs and LAs comprise the TMDL, or the maximum discharge of pollutant in the basin that will allow the water quality standard to be met. A necessary component of this process is the identification of all loads and an assessment of the assimilative capacities of the waterbody, in relation to the water quality standards to be met.

Point source discharge permits are based on the WLAs for the basin. However, provisions of the Act allow for regulators to consider the relative costs of control when issuing discharge permits. The law states that 'if Best Management Practices (BMPs) or other non-point source pollution controls make more stringent load allocations practicable, then wasteload allocations (point source controls) can be made less stringent. Thus the TMDL process provides for non-point source control tradeoffs' (Bartfeld, 1993, p. 73). However, the TMDL process does make non-point sources legally responsible for meeting LAs, as the NPDES permits do for WLAs.

North Carolina has identified several basins as being Nutrient Sensitive Waters (NSW) where the TMDL process is being applied, one being the Tar-Pamlico Basin. Nutrient enrichment had led to massive algal blooms and a degradation of commercial and recreational fishing. The Tar-Pamlico programme provides good examples of several problems facing existing point/non-point trading programmes. The largest point source polluters in this area were formed into an association and traded as a group (to reduce transactions costs) at a predetermined price. Members of the association could purchase nitrogen reduction allowances by contributing to the North Carolina Agricultural Cost Share Program at a fixed price of $56 kg^{-1} (this price has recently been reduced to $29 kg^{-1}). The state would then handle the task of getting agricultural producers to participate in the programme and deciding how much reduction would be achieved by alternative farming practices. However, the fixed price was based on average control costs, thus reducing the potential benefits that would have been obtained through margin pricing (Hoag and Hughes-Popp, 1997). Also, the requirement of a 2:1 trading ratio may have increased the cost of a trade to levels that would have been unattractive to point sources. Initial loading reduction goals for the programme were met by the point sources through process changes at a cost of less than $56 kg^{-1}. Finally, the programme is hampered by a lack of generally applicable models or data linking land-use practices to water quality effects (Hoag and Hughes-Popp, 1997).

Comprehensive non-point source programme – Kentucky

In 1994, Kentucky passed the Kentucky Agriculture Water Quality Act (SB 241; codified in Ky. Rev. Stat. 224.71). Its main goal is to protect surface and groundwater resources from pollution that may result from agricultural activities, including sediment and agricultural chemicals. As such, it is one of the few comprehensive water quality protection laws in the nation. The Act requires all land users with 10 or more acres (4 ha or more) to develop and implement a water quality plan based upon guidance from the Statewide Water Quality Plan. Guidance is in the form of 58 approved best management practices. A farmer will select applicable BMPs to be included in his plan. Education, technical assistance and financial assistance (conditional on availability of funds) will be available for the development and implementation of the plans. Landowners will have 5 years to implement their plans.

In cases of unique local water quality problems where agriculture has been identified through monitoring as a major contributor, the law provides for creation of water priority protection regions. A regional water quality plan would be developed for the priority region. Modifications to individual plans may be required to meet a protection region's water quality goals. Technical and financial assistance will be made available to assist landowners in modifying their plans. Land users must comply with the regional water quality plan to receive assistance.

If a watershed is still impaired after 5 years, all operations will be checked for approved BMPs. Those farmers not using the necessary practices will be given another opportunity to adopt them. Assistance will be again provided to make the necessary changes. However, failure to take protective measures may result in a 'bad actor' label, making the landowner subject to enforcement action, including fines of not more than $1000.

Animal waste

Nutrients from livestock manure are an increasing concern across the US, given the recent trend towards larger, more specialized beef, swine and poultry operations. Approximately 450,000 operations nationwide confine or concentrate animals (USEPA, 1998b). Of these, about 6600 have more than 1000 animal units, and are defined under the Clean Water Act as Concentrated Animal Feeding Operations, or CAFOs. Such operations must handle large amounts of animal waste USDA and EPA (1999). There are two sources of water quality problems from CAFOs: (i) they require large and sophisticated manure handling and storage systems, which have at times failed, with serious localized consequences; and (ii) CAFOs tend to lack sufficient cropland on which manure can be spread without exceeding the plant nutrient needs (Letson and Gollehon, 1996). Excess application of waste (providing more nutrients than plants need) can lead to non-point pollution problems. There are many instances over the past several years where animal waste storage lagoons have broken or leaked, or where excess applications to the land have adversely affected water quality (NRDC, 1998).

The CAFOs are treated under the Clean Water Act as point sources. They therefore need an NPDES permit in order to operate. The standard permit states that all runoff from the site resulting from a storm of less intensity than the 25-year/24-hour storm be collected and stored. However, the traditional permit does not cover application of waste on cropland or other land. So, when storage lagoons are pumped out, the material can be spread on fields at rates far beyond plant needs, resulting in the potential for a non-point problem.

A number of states have started to pass laws that address the problem of what happens to animal waste once it leaves the CAFO. An approach that is becoming more common is to require a nutrient management plan for the application of the waste. The plan can be made one of the requirements of the NPDES permit, or it can be required through a separate law. In some cases, the state requires a nutrient management plan even if the manure is sold or given to another landowner. Currently, 23 states require some form of nutrient management plan for at least some classes of animal operations.

An example is Illinois. Land application of animal waste in Illinois is allowed subject to regulations established by the Livestock Management Facilities Act, which became effective in 1996 (Il. Admin. Code Title 35, section 505). Under the regulations, livestock management facilities with 1000 animal units must prepare and maintain a waste management plan. Operations with 7000 or more animal units must submit this plan to the Department of Environmental Protection for approval. The plan must demonstrate that the maximum nitrogen application rate to obtain optimum crop yields is not exceeded. Required in the plan is information on where the waste will be applied, how it will be applied, application rates, nitrogen carryover from previous crops, and cropping and yield histories of the fields receiving waste. Enforcement is through citizen complaint or inspection.

Issue of Enforcement

An issue in many state non-point source pollution control laws is whether they are providing adequate incentives for landowners to adopt BMPs. The incentive provided by the technology-based regulatory framework adopted by most states is the threat of being caught and penalized for not using required BMPs. Many of these programmes or laws rely on citizen complaint for identifying producers not in compliance, particularly for those programmes not targeted to a specific problem in a specific area. The problem with non-point source pollution is that the origin of a water quality problem cannot be readily identified (see Chapter 1). If the first sign that a producer is discharging pollutants to waters is a fish kill or some other visible water impairment severe enough to spur a citizen complaint, the damage has been done and it still may not be possible to identify the culprit. An alternative approach is site inspections, but these are expensive, would probably be unpopular with landowners, and would still not guarantee that those responsible for a water quality impairment are identified and penalized.

Future Directions: the Clean Water Action Plan

Although the CWA has resulted in a great number of successes, many water quality problems remain – especially those related to non-point sources (USEPA, 1998c). Instead of waiting for the next reauthorization of the CWA (which has been delayed for 6 years), the White House ordered USEPA and USDA jointly to develop a Clean Water Action Plan (CWAP) with assistance from other federal agencies. The CWAP was released in February 1998. It is an ambitious proposal that lays out a fundamental shift in water quality policy to emphasize control of non-point sources of pollution, especially sources of polluted runoff, using existing laws and authorities for more complete water quality protection.

Many of the 111 action items outlined in the CWAP are directed towards polluted runoff, or non-point source pollution. Polluted runoff (especially nitrogen and phosphorus) is an important source of remaining problems, with agriculture as the largest single contributor. Since agricultural operations are major sources of polluted runoff, programmes developed to carry out CWAP initiatives will likely place greater pressure on farmers in impaired watersheds to address runoff problems. Whether the programmes will use voluntary approaches similar to current programmes or new, innovative approaches has yet to be determined.

One of the earliest actions completed was the development of a strategy for addressing the water quality and public health impacts associated with animal feeding operations. USDA and USEPA released the Unified National Strategy for Animal Feeding Operations on 9 March 1999. The major goal for the strategy is that all animal feeding operations develop and implement technically sound, economically feasible and site-specific Comprehensive Nutrient Management Plans (CNMPs) that address how animal waste is managed both on the site and when applied to land. The contents of the CNMPs and the mechanisms for getting producers to develop and implement these plans will be addressed in guidance and rules to be developed by USEPA and USDA. USEPA will also revise the permitting rules for confined animal feeding operations to clarify which operations will be required to obtain an NPDES permit.

The CWAP acknowledges USDA's experience in working with farmers on a watershed basis. Specifically, USDA will have a role in helping states to develop watershed protection goals and water quality protection strategies, along with USEPA. In addition, USDA will be a major source of education, technical assistance and financial assistance to landowners developing comprehensive management plans to protect water quality. Current USDA programmes such as the Environmental Quality Incentive Program, Conservation Reserve Program, Wetland Reserve Program and Wildlife Habitat Incentive Program can all provide incentives to farmers for addressing water quality concerns. The CWAP proposes increased funding for USDA to support water quality efforts.

Summary and Conclusions

The current institutional structure for protecting water quality in the US is weighted heavily towards dealing with pollution from point sources, particularly at the federal level. Responsibility for non-point source pollution control has been given to the states, with the federal government providing scientific, technical and financial support. Point sources are controlled through a centralized command-and-control programme based on technology and performance standards, while non-point sources are largely addressed through voluntary incentives provided by federal (USDA and USEPA) and state programmes. This dichotomy of treatment between point and non-point sources has been criticized for preventing a more efficient protection of water quality (Freeman, 1994; Davies and Mazurek, 1998). The traditional focus on point sources is becoming less effective at eliminating the major threats to water quality that are increasingly being attributed to non-point sources.

The characteristics of non-point source pollution have a major influence on the current institutional structure for controlling pollution from non-point sources. The large variability in characteristics of non-point source pollution over geographical space led to a programme that emphasizes local control. The inability to identify the sources of impairments from non-point sources, and the sheer number of individual sources, has led to an emphasis at both the federal and state levels on voluntary approaches that support best management practices and other land-use changes through education, technical assistance, financial assistance and research.

Some states have developed non-point source programmes that rely on enforceable mechanisms for requiring the adoption of best management practices. A wide variety of institutional structures have been put in place, tailored to specific concerns of each state. Most state policies rely on technology-based standards, in the form of approved best management practices, largely because performance-based approaches are impractical or impossible for non-point sources. In the case of Florida, where it is possible to link water quality to practices, performance-based approaches are being tried. Enforcement is often based on citizen suit rather than inspection and monitoring, reflecting the potentially high administration costs of such programmes, where there are numerous potential sources. The wide variation in degree of effort across the states is also partly indicative of the availability of fiscal and technical resources for implementing non-point policies. None of the federal or state programmes accounts for the stochastic nature of non-point source pollution.

If the Clean Water Action Plan released by the Clinton administration is any indication, there will be a greater emphasis on achieving specific water quality goals in the future (White House). The plan targets non-point source pollution for greater control, calls for establishing specific water quality standards for achieving the 'fishable' and 'swimmable' goals of the Clean Water Act at the watershed level, and calls for the use of enforceable mechanisms to ensure that appropriate management practices are adopted (White House).

Such approaches would put a tremendous burden on regulators to identify what the appropriate management practices are for meeting water quality standards, and then to enforce them.

Non-point source pollution is difficult to regulate and to control, as it usually results from numerous geographically dispersed sources, each emitting small amounts of pollutants. An effective non-point policy will have to influence many actors to reduce relatively small amounts of pollution that cannot be observed. A policy framework such as that laid out by Segerson (1999) is one possible approach for achieving more cost-effective control. This approach uses a variety of tools, such as subsidies, education and performance standards in a 'trigger' policy framework to get farmers to meet ambient water quality standards. Research on the linkages between management practices and water quality and the development of water quality models would greatly assist farmers and resource managers in linking land-use activities to water quality.

No general statement can be made about which policy instrument currently being used gives the most efficient or cost-effective control of non-point source pollution. The characteristics of non-point source problems vary tremendously across the country. The choice of policy to control non-point source problems depends on the nature of the environmental quality problem, on the information available to the administering agency on the linkages between farming activities and environmental quality, on farm economics, and on societal decisions about who should bear the costs of control. An approach based on state and locally developed watershed-level control plans that allow a variety of policy tools to be used, including both carrots and sticks, probably provides the greatest opportunities for cost-effective control.

Chapter 6

Policy on Agricultural Pollution in the European Union

NICK HANLEY

Department of Economics, University of Glasgow,
Glasgow G12 8RT, UK

Policy at the EU Level

Increasingly, the focus of environmental policy formation in Europe is moving away from nation states and towards the EU itself and its executive arm, the European Commission. Policy passed at the EU level, in the form of directives and regulations, establishes targets and requirements which national policy must subsequently meet (although the lag between a directive coming into force and national policy innovations to fulfil its requirements varies a lot across the EU). Directives and regulations themselves follow broad principles as set out in the EU Environmental Frameworks, and through the Council of Ministers. Thus, for example, the 'polluter pays' principle lies behind many directives. The subsidiarity principle dictates that the fine-tuning of legislation should be decided at the national level, rather than the EU level, though what this national legislation must achieve can be very detailed. Thus, whilst EU directives set the direction of policy change, and in some cases the targets that must be met (e.g. in terms of water quality), the details on actually achieving these targets, or in adapting to new directions, are left up to member states.

With regard to pollution from farming, the most important EU measures are the Drinking Water Directive (75/440 and 80/778), the Nitrates Directive (91/676) and the Agri-Environmental Regulation (92/2078). The *Drinking Water Directive* was important because it established maximum permissible concentration levels of polluting substances in potable water supplies. Nitrate was one of the substances included, and an upper limit of 50 mg l^{-1} was set. This immediately created problems for some EU countries,

© CAB *International* 2001. *Environmental Policies for Agricultural*
Pollution Control (eds J.S. Shortle and D.G. Abler)

since water supplies in certain farming areas were found to exceed this level. The UK, for example, had to apply to the European Commission for derogations (temporary exemptions) from the directive with regard to water supplies in the English Midlands and East Anglia.

The *Nitrates Directive* brought about the need for action in many EU countries. The directive contains three main provisions:

- Countries must monitor all waterbodies to identify areas where water quality is threatened by nitrate pollution from agriculture. These threats are defined in terms of actual or potential eutrophication problems, and actual or potential exceeding of the 50 mg l^{-1} limit for nitrates in drinking water supplies. Affected areas are to be designated as Nitrate Vulnerable Zones (NVZs).
- For all NVZs, member states must produce action plans which target animal manure and inorganic fertilizer applications to remove these threats. These action plans should be enforceable.
- For all other areas of the country, member states must produce voluntary codes of good agricultural practice, which cover stocking rates, application rates, timing of application and other relevant issues.

Action plans in NVZs must include maximum allowable application rates for minerals from animal manures, for example 170 kg N ha^{-1}, unless other actions are taken which compensate for a less restricted rate being allowed. Action plans also need to include provisions for periods of the year when application is prohibited, storage of animal manures and maximum application rates for inorganic fertilizers.

The *Agri-Environmental Regulation* is mainly important with respect to the protection of wildlife and landscape with regard to farming. It establishes some general principles for the design of national policies designed to protect wildlife/landscape by making payments to farmers. The way in which the regulation has been applied nationally varies greatly within the EU (see, for example, Dabbert *et al.*, 1998), particularly with regard to the extent to which regulation of farming activities is permitted. There are, however, some implications for water quality – for example, where measures are taken to encourage organic farming; or where habitat protection impacts on bankside vegetation (see the section on the UK below). With regard to pesticide use, *Council Directive 91/414* from 1991 demanded a re-registration of all pesticides licensed before 1993. Again, this has necessitated action at the country level to meet EU requirements (see the section on Scandinavia below).

Trends in EU-wide agri-environmental policy also need to be seen in terms of the general direction of reform in the Common Agricultural Policy (CAP). Historically, the CAP operated by setting guaranteed prices for almost all farm outputs above market-clearing levels, and maintained these through a system of export subsidies, intervention buying and tariffs. However, the very high cost of the CAP to taxpayers and consumers led to pressure for reform in the 1980s and 1990s, which has resulted in a redirection of support away from a focus on prices. Support is now increasingly being arranged through area-based payment schemes, although headage payments for livestock in less favoured (i.e. marginal) areas is still maintained. The expansion of agri-environmental schemes is seen as a part of this general reform of the CAP, in that it will allow farmers to earn income from 'producing' environmental goods. So far, however, total spending on such schemes is still a very small fraction of total CAP spending, at around 2% (Hanley et al., 1999).

Policy at the Individual Country Level

United Kingdom

Within the UK most government attention has historically been given to agriculture's impact on wildlife and landscape, rather than on water quality. This is due to the fact that whilst the impacts of changes in farming practice on wildlife and landscape have been widespread and very noticeable, water quality impacts have been somewhat localized and not so noticeable. The early evolution of environmental policy focused heavily on industrial sources of pollution, and on civic responsibilities for waste collection and treatment. Early legislation on the countryside was more focused on recreation (e.g. the National Parks and Access to the Countryside Act, 1949; the Countryside Act, 1968). In 1974, the Wildlife and Countryside Act allowed for the safeguarding of Sites of Special Scientific Interest through bilaterally negotiated management agreements, funded by the Nature Conservancy Council. Interestingly, up until the mid-1980s, there was no acknowledgement by the Ministry of Agriculture, Fisheries and Food (MAFF) that farmers had any other than a good impact on rural amenities, and certainly no suggestion that agriculture budgets should be used to pay for environmental improvements. This all changed in response to the 1985 Structures Regulation from the EU, which for the first time allowed farm ministry budgets to be expended on environmental protection. This, reinforced by the Agri-Environmental Regulation of 1992, gave rise eventually to a large number of agri-environmental policies part-funded by MAFF, all expressly concerned with the supply of environmental goods by farmers (Hanley et al., 1999).

These agri-environmental policies all follow a standard UK approach, which might be termed the 'management agreement' model. This same approach in fact underlay the earlier Countryside Act and the Wildlife and Countryside Act, and assumes that farmers have the implicit property rights over environmental services provided by their land (see Box 6.1). Regulation is thus not favoured; rather, payments are offered for voluntary participation in schemes, the success of which is typically measured by participation rates (Hanley et al., 1998a, 1999).

Box 6.1. The UK Management Agreement model.

In the UK, almost all agri-environmental policy follows the following principles:
- Farmers are assumed to have the property rights over all services provided by their land, including environmental services.
- Therefore, farmers cannot be forced either to abstain from actions which damage the environment, or to undertake actions which favour the environment.
- By a similar logic, negative environmental impacts originating from farming should not be taxed.
- A system of subsidies in return for voluntary participation in schemes is favoured.
- These subsidies are offered as part of a contract whereby the farmer agrees either to undertake certain actions (e.g. managing hedgerows, cutting stocking rates, maintaining low-level grazing on heathland) or to abstain from certain actions (e.g. draining wetlands, felling farm woodlands).
- Payments can be uniform within a given area, uniform across the whole country, or tiered as part of a menu of payments and obligations.
- Success of the scheme is typically measured by participation rates.

Whilst most agri-environmental policy is aimed at wildlife and landscape protection, farming's impact on water quality is nevertheless given some attention. But what are these impacts? They include: (i) nitrate and phosphate pollution of rivers, lakes and streams; (ii) pesticide drift; (iii) spillages of silage effluent from storage areas (silage effluent is 200 times more polluting than domestic sewage in terms of BOD[1] impacts); (iv) pollution from the storage, spreading and deposition of livestock wastes (about 200 million tonnes a year are disposed of); (v) accidental spillages of oil or diesel from storage tanks; and (vi) runoff water from vegetable washing plants.

Problems from silage effluent and manure spreading are concentrated in dairy farming areas, predominantly in the western half of the country. Soil erosion is not seen as a significant threat to water quality in the UK. Overall, water pollution from farming seems to be declining. In 1988 the National Rivers Authority recorded 4141 incidents of water pollution due to farming; this figure had fallen to 2050 by 1998. These figures, however, only relate to easy-to-notice pollution events, such as fish kills, and not to all agricultural sources of water pollution. Pesticide use is also seen as an environmental threat. There are around 450 different active ingredients licensed for use in the UK. In 1995, the National Rivers Authority monitored concentrations of 160 different pesticides at 2500 sites (DETR, 1997). At 8% of these sites, actual concentrations were found to exceed maximum desirable levels, known as Environmental Quality Standards. This was a slight increase over the previous year. The most commonly occurring pesticides in groundwater were found to be atrazine and isoproturon.

Early legislation on water pollution from farming was noticeable mainly by its absence. MAFF issued 'best practice' guidelines to farmers, which showed them how to minimize risks of water pollution – for example, by not spreading manure on frozen fields, or by not spraying pesticides on windy days. The main aim of this guidance was to prevent pollution incidents from occurring. However, failure to abide by such codes was not a legal offence. These guidance notes have continued to be updated and now cover water pollution, air pollution and soil conservation. Farmers can be prosecuted for allowing silage effluent spills to pollute streams under the Water Resources Act of 1991, and recent changes in legislation also allow the regulator to insist on farmers taking adequate precautions to prevent spillages of all substances that might pollute watercourses (under the Silage, Slurry and Fuel Oil Regulations, 1991). Local authorities can attempt to take actions against offensive smells (e.g. from pig units) under the Environmental Protection Act, 1990.

The major piece of legislation on water pollution from farming is the Nitrate Sensitive Areas (NSA) scheme. The NSA scheme was introduced in 1990 in response to the EU Drinking Water Directive, which compelled member states to take actions to prevent nitrate levels in drinking water rising above the World Health Organization limit of 50 mg l^{-1}. There are now 32 NSAs designated in England and Wales, in areas where farm practices, hydrology and soil type led to nitrate standards being either violated or threatened with violation. Farmers in these areas, which account for 35,000 ha of farmland, can opt into the scheme by agreeing to restrictions on their farming practices in return for standard payments per hectare. These payments vary across and within NSAs. Farmers may enter part or all of their farms, and sign agreements lasting 5 years. Prescriptions include: (i) converting arable land to unfertilized, ungrazed grass; (ii) converting arable land to grass fertilized up to a fixed limit (e.g. 150 kg N year^{-1}); and (iii) low-nitrogen arable cropping, e.g. up to 150 kg N year^{-1}, with no potatoes or brassica production allowed. All contracts include requirements to maintain trees, hedges, walls and historical features on contracted land; this is not aimed at reducing N pollution but at generating other environmental benefits.

Payments range from £625 to £65 ha^{-1}, depending on area of the country and extent of restrictions agreed to. Across all NSAs, about 58% of eligible land had been entered into the scheme under the basic scheme, with a further 21% under the premium arable scheme by 1998. Monitoring has shown that the scheme has resulted in falling nitrate runoff levels in affected areas. The scheme runs until 2003 but was closed to new entrants in 1998. There is some evidence that the value of environmental benefits resulting from the NSA scheme exceeds its costs by a considerable margin (Stewart *et al.*, 1997).

In response to the EU Nitrates Directive passed in 1991, the UK introduced a separate NVZ scheme in 1998. This directive, as will be recalled, is more wide ranging than the Drinking Water Directive, since it is also concerned with ecological definitions of nitrate pollution, such as eutrophication, and not just with human health implications to do with N levels in drinking water. The main difference is that farmers in NVZs are not compensated for losses involved in meeting mandatory Action Programme measures, mainly connected with the rate and timing of inorganic and organic fertilizers, as the government believes these requirements to be consistent with 'good agricultural practice'. There are 68 NVZs in the UK, covering some 600,000 ha.

A pesticide tax

Devising a tax on pesticides to reduce their environmental impacts is a hard task. Environmental impacts depend on not only the chemical composition of products, but also how they are used, how often they are used and where they are used. This suggests a tax which is differentiated by constituents, and which is backed up by regulations on how pesticides are used (DETR, 1997). Governments also need to be aware that if one pesticide is taxed, farmers will substitute others, yet these replacements might impose greater damages. Pesticide use has traditionally been managed in the UK through the licensing of products by government, and by voluntary codes of practice on how pesticides should be used to minimize environmental damage and health impacts on humans. Licensing is carried out through an informal balancing of benefits and risks of use, with the constraint that no 'unacceptable risk' to human health and ecosystems is expected. However, licensing imposes no limits on the amount or extent of use, or on the manner of use. It also gives farmers no incentive to cut back on pesticides use below privately optimal levels.

The UK has thus been exploring a tax on pesticides, imposed as a product charge, which would aim to reduce absolute quantities of use and encourage substitution to less harmful products. This latter objective would require a banded tax system, with products being grouped on some measure of expected damage. The tax would be based on estimated marginal costs of pesticide reduction, rather than estimated damage costs. A feasibility study judged that such a tax would also encourage the take-up of integrated pest management systems, which inherently use less pesticides. The price elasticity of demand for herbicides was estimated as lying between -0.3 and -0.7. It was estimated that a 50% tax rate on sales price would reduce farm incomes, on average, by about 2% and generate tax revenue of £70–80 million per year in the short run.

Other policy initiatives

Policy on the environmental impacts of farming in the UK is dominated by agri-environmental schemes that pay farmers to produce environmental goods such as pleasing landscapes and high quality wildlife habitats. However, some of these policies have impacts on agricultural pollution. Thus, the Organic Aid scheme, which encourages farmers to switch production methods to certified organic ones through a system of per-hectare payments, is also expected to reduce pesticide use and thus pesticide pollution. Water quality impacts with regard to nutrients and BOD could be either positive or negative, since reductions in inorganic fertilizer use are offset with increases in organic fertilizers, which can have equally bad impacts on nutrient levels and which have a much higher BOD impact. The Environmentally Sensitive Areas scheme may have beneficial impacts on water quality where, for example, buffer strips are created in river valley grasslands.

Scandinavia[2]

Non-point pollution problems linked to agriculture in Scandinavia include: (i) eutrophication of coastal waters (e.g. in the Baltic and Kattegat) due to runoff of nitrates and phosphates from farmland (and forests); (ii) nitrate and pesticide residues in groundwater; (iii) water quality in inland lakes and rivers (though reductions in point source emissions, especially sewage, have meant that overall water quality has risen); and (iv) concerns relating to pesticide use.

Pollution from nitrates and phosphates

Nitrate and phosphate pollution from agriculture originates from the use of both inorganic fertilizers and animal manure. Historically, nitrogen fertilizer use rose in Denmark from 1970 until about 1990. Use in Sweden and Finland was roughly constant over this period. Phosphate use has fallen in all three countries during this period, from about 460,000 tonnes to 190,000 tonnes (Brannlund and Kristrom, 1999). In response to concerns over nutrient enrichment, Norway, Sweden, Denmark and Finland have all introduced policies to control fertilizer use. A mixture of command-and-control regulatory initiatives and economic incentives have been employed. In Denmark, for example, the emphasis has been on regulation, in terms of maximum allowable application rates of manure, seasonal bans on application, and minimum storage requirements (Schou, 1997).

Fertilizer taxes have also been employed. Sweden, for instance, currently imposes a tax of SEK 1.80 kg^{-1} on the nitrogen content of fertilizers. Finland had a tax on nitrogen fertilizer until 1994 but phased this out on joining the EU, as it feared for the competitiveness of its small farm sector. Norway retains a tax on nitrogen and a tax on phosphorus, at rates equal to SEK 1.20 and SEK 2.30 kg^{-1}, respectively.

As Brannlund and Kristrom (1999) pointed out, a uniform tax rate on fertilizers is unlikely be economically efficient, if the objective is cost-effective reduction of actual nitrate pollution. To illustrate this, they computed the environmental impact coefficients for nitrate loading reaching coastal waters for a number of different regions of origin in Sweden (Table 6.1). The authors pointed out that the tax rate that farmers pay per unit of N applied should be highest where impacts are great, and lowest where impacts are small. This is what the final column of Table 6.1 shows: the tax rate should be highest in The Sound region and lowest in the Bothnian Bay region. However, since environmental impacts will also vary within regions, even this differentiated tax is not fully efficient.

Table 6.1. Impact coefficients and efficient tax rates for a fertilizer tax in Sweden.

Region	Impact coefficient	Efficient tax (SEK kg^{-1})
Bothnian Bay	0.03	0.45
Bothnian Sea	0.19	2.85
Malar region	0.03	0.45
Baltic Sea region	0.07	1.05
The Sound	0.21	3.15
Kattegatt	0.10	1.50
Skagerack	0.17	2.55

Source: Brannlund and Kristrom (1999, p. 26).

A complication regarding eutrophication impacts in parts of Scandinavia is that many countries contribute to them. For example, eutrophication in the Baltic Sea is caused by runoff of nitrogen and phosphorus from Sweden, Denmark, Finland, Germany, Poland, Estonia, Latvia and Lithuania (Table 6.2). Effective action thus requires coordination between countries, but the public-good nature of subsequent improvements implies that countries will behave strategically and not cooperate fully. Table 6.2 also shows the marginal impacts of nitrogen use in the different countries. Nitrate taxes should be highest where the marginal benefits of reducing nitrogen use are greatest in terms of reduced total loading. This implies that Denmark should have the highest tax rate and the Eastern group of countries the lowest. In fact, only Sweden amongst this group of countries currently taxes nitrogen use.

Table 6.2. Marginal impacts of nitrate use across countries.

Country	Marginal load	Percentage load of nitrogen from arable land
Sweden	0.09	11.9
Finland	0.11	11.2
Denmark	0.15	14.1
Germany	0.13	25.3
Poland	0.05	40.0
Eastern group	0.03	13.0

Source: Brannlund and Kristrom (1999, p. 43).

Pesticides

All of the Nordic countries impose a tax on pesticides. The case of Denmark in particular is an interesting one. Denmark had originally taken a stand against the use of economic incentives in environmental protection. The Minister of the Environment stated in 1973 (cited in Andersen, 1999) that such instruments were undesirable because 'those who can afford it will be allowed to pollute, and those who cannot afford it will not be allowed. We do not want to bring class policy into environmental policy.' However, by the end of the 1990s Denmark had introduced an extensive programme of eco-taxes, aimed at both changing behaviour and allowing revenue-raising to be partly switched away from distortionary taxes. For example, the introduction of a charge on dumping non-hazardous wastes increased recycling rates from 35% to 61%, whilst increased eco-tax revenues of DKK 12.2 billion allowed an equivalent reduction in income taxes (OECD, 1997).

The origins of the Danish pesticides tax date back to the 1986 Pesticides Action Program, which aimed to remove all pesticides identified as unacceptably hazardous and, in addition, to reduce usage of the remaining pesticides by 50%.[3] This plan certainly achieved some success: half the active ingredients previously approved for sale in Denmark were removed from the market, and the quantity of active ingredients applied dropped by more than 40% (Dubgaard, 1999). A further result was that 105 of the 218 active ingredients previously approved for sale in Denmark were withdrawn by suppliers. Even with this reduced usage of less hazardous material, environmental groups argued that excessive environmental risks persisted. Accordingly, a pesticides tax was introduced in 1995 as a way of reinforcing existing regulatory initiatives. The tax was specified as an *ad valorem* tax, set at rates of 37% on insecticides and 15% on all other pesticides, and these tax rates were doubled in 1998. However, as Dubgaard noted, use of an *ad valorem* tax implies that the most expensive pesticides are those imposing the highest damages. This is quite clearly not so, as some of the older (and therefore cheaper) substances are associated with high levels of environmental risk. An *ad valorem* tax could thus lead to substitution away from more expensive, lower damage products to cheaper, higher damage products. As a solution, Dubgaard argued in favour of changing the tax system to one weighted by environmental impact, using an appropriate indicator.

The Netherlands

The Netherlands is one of the most densely populated countries in the world. Much intensive livestock production is located there, leading to environmental problems relating to livestock wastes; in addition, the large horticultural industry is a heavy user of pesticides. This section focuses on how The Netherlands responded to the EU Nitrates Directive (Frederiksen, 1997). Due to soil characteristics and the nature of farming activities, the Dutch government decided in 1995 to designate the whole of the country as a Nitrogen Vulnerable Zone. This clearly had major implications for agriculture. The principal means of measuring compliance with action plan targets is a farm-level mineral accounting system, which uses a mass balance approach to estimate for *each farm* the level of nitrate loss in any year, where nitrate loss is the excess of inputs over crop take-up. The system will thus allow the government to quantify the environmental loading in any year from nitrates across the whole country, and to relate localized eutrophication or drinking water quality problems to reported nitrate losses. This accounting system is to be gradually phased in, with intensive livestock producers being the first to use the system from 1998 onwards.

Action plan requirements can also be spelt out. A range of fertilizer applica-tion limits are set; for example, a maximum of 200 kg P_2O_5 ha^{-1} can be applied on grassland. Application of animal manures is forbidden between 1 September and 1 February on soils vulnerable to leaching. Calculated farm manure sur-pluses will be taxed, based on compulsory construction of manure balance sheets for each farm. A manure trading system is also in operation, whereby farmers buy and sell the rights to spread manure, thus allowing farmers in 'spare capacity' areas to sell spreading rights to farmers in areas where no further appli-cations are allowed. These rights are defined on a comparison of actual livestock stocking rates on a given farm with maximum allowable spreading rates. The trading system has an inbuilt decreasing supply mechanism, whereby the entitle-ment is cut by 25% each time a trade occurs that is not tied to land. Finally, the government has supported industrial reprocessing plants for manure surpluses. Helming (1997) produced some estimates of the implications of these policy changes on farm incomes and environmental quality in The Netherlands.

Lessons and Remaining Challenges

Environmental objectives are increasingly becoming an expected component of agricultural policy reform in the EU. This is a dramatic change from the situation even 15 years ago. The 1985 Structures Regulation marked a turn-ing point in the sense that: (i) it recognized that agriculture had impacts on the environment and (ii) that agricultural ministry budgets could be used to support policy initiatives aimed at improving this aspect of agricultural performance. As we have seen, the way in which EU policy works is by the EU setting targets and/or frameworks, but national policy setting the details on how targets are to be achieved or frameworks implemented.

All EU countries now regulate the environmental impacts of farming to a degree. Intervention, we have seen, takes three main types: (i) environmental taxation in some cases (e.g. the Swedish nitrates tax, the Danish pesticides tax, the Dutch manure levy); (ii) voluntary sign-up programmes, whereby farmers agree to certain management restrictions/requirement in return for payments; and (iii) regulation, such as stocking limits, spreading limits, restrictions on the timing of fertilizer applications, pesticide regulation, or due-care standards over storage of potentially polluting substances. Voluntary sign-up schemes dominate the case of wildlife and landscape pol-icy, where on the whole society sees farmers as producing public goods which need to be paid for (Hanley et al., 1998a). Environmental taxation and regula-tory measures dominate in the case of environmental bads, such as water pol-lution. Here, society seems to take the view that on the whole farmers should pay to clean up the environment to socially desirable levels. This difference in the implicit allocation of property rights is interesting, and has been alluded to elsewhere (Bromley and Hodge, 1992). It shows that what is defined as a public bad or public good influences the choice of policy instrument.

What lessons can be learnt from existing policy? And how can policy design be improved? Several themes seem to be important. Firstly, there is a need to improve coordination between agri-environmental policy and agricultural policy, and indeed other rural policies. Article 130R of the Maastricht Treaty calls for the greater integration of environmental and agricultural policies, yet we still find the CAP and agri-environmental schemes pulling in opposite directions. For instance, Environmentally Sensitive Areas in upland regions of the UK frequently suffer from over-grazing by sheep, which reduces their ecological quality. ESA payments thus encourage farmers to reduce grazing pressures by reducing sheep numbers, but the system of Hill Livestock Compensatory Allowances under the CAP encourages farmers to keep more sheep, by offering a headage payment per breeding ewe. This conflict, of which there are many other examples, both increases the apparent costs of conservation and reduces take-up rates.

Cross-compliance has often been suggested as a means of integrating environmental goals with the broader goals of farm policy. Many different forms of cross-compliance exist, such as 'red ticket' approaches, where farmers lose all entitlements to support payments if they fail to meet environmental standards, 'green ticket' approaches, where farmers get higher levels of support if they meet these standards, and 'pink ticket' approaches, where farmers lose part of their entitlements if they violate environmental conditions (Batie and Sappington, 1986; Baldock and Mitchell, 1995). The difficulties of implementing credible cross-compliance schemes are well known (see, for example, Spash and Falconer, 1997); however, there has been a limited but increasing use of cross-compliance in the EU (for example, environmental requirements for set-aside payments) and in Norway (where environmental conditions are attached to arable support payments). The UK government, for one, has called for the increasing use of environmental cross-compliance within the CAP in future, citing the example of environmental conditions being imposed on the arable area payment scheme, and on headage payments for livestock (DETR, 1997).

Secondly, there is a need for more targeting of agri-environmental programmes. Payments for the production of environmental goods from rural land management should ideally be targeted where these are most valued. For example, only 10% of the land area of Finland is arable land, and much of this is highly valued for its ecological and cultural values. Yet Finnish applications of the Agri-Environmental Regulation (92/2078) have focused on nutrient run-off problems, rather than on conserving the biodiversity and landscape values of this land (Sumelius, 1997). In a different vein, efforts to protect habitats may be hampered because payment schemes are administered on an area-wide basis, rather than a habitat-specific basis.

Thirdly, uniform payment rates are inefficient if opportunity costs of meeting scheme requirements vary across farmers: there is ample evidence that this is indeed the case (Hanley *et al.*, 1998a). Next, many voluntary payment schemes are set up in terms of required management actions (e.g. a reduction in stocking rates, a reduction in fertilizer use), but links between management actions and environmental effects are often uncertain. Better then, some would argue, to target policy at environmental outcomes. As we know, this is a problem, due to the non-point nature of much agricultural pollution. Environmental outcome targeting may thus be better suited to policies aimed at the production of wildlife and landscape (since the identification problem is easier), whilst management action targeting is better suited to policies aimed at reducing environmental bads such as pollution.

Finally, it is apparent that very few environmental policies within agriculture in the EU are subject to any kind of economic efficiency criteria. These include cost-effectiveness and Pareto efficiency. With regard to the former, there has been little official recognition of the role that economic instruments can play in bringing about cost-effective solutions to externality problems. Where economic instruments have been used, the main purpose has been either as a gesture towards the 'polluter pays' principle (e.g. nutrient taxes in Sweden and Finland), or as a means of improving target achievement (e.g. the Danish pesticides tax). To some degree, this is a function of the difficulties in designing taxes and permit markets for non-point pollutants, but mainly one suspects it arises from a failure on the part of economists to sell their case well enough. The direction of change is, however, promising, with The Netherlands planning to place more emphasis on the market and on taxes to achieve its responsibilities under the Nitrates Directive, and with the general use of economic instruments in environmental policy rising throughout the EU.

With regard to Pareto efficiency, economists would prefer to see more governments applying cost–benefit analysis criteria to the control of agricultural externalities. Whilst there is some use of this in the context of wildlife and landscape policy in the UK (as summarized in Hanley *et al.*, 1999), and whilst similar policies have also been appraised in this way in France, Sweden and Norway, there are a great many more instances where cost–benefit criteria are not applied. This may be leading to some big policy mistakes. Are the costs of meeting the Nitrates Directive in the EU really warranted by the expected environmental and health benefits? Should a much greater level of resources be put into agri-environmental schemes under regulation 92/2078? Are the costs of the Danish pesticides action plan too high? At the present, we just do not know the answer to these important questions.

Endnotes

1. BOD is biological oxygen demand, a common measure of the polluting potential of organic substances.
2. For a useful account of point source pollution control in Denmark, see Andersen (1999).
3. This is defined as a 50% reduction in the volume of active ingredients applied as well as a 50% reduction in the annual number of treatments per unit area (Dubgaard, 1999).

Chapter 7

Decomposing the Effects of Trade on the Environment

DAVID G. ABLER AND JAMES S. SHORTLE

Department of Agricultural Economics and Rural Sociology,
Pennsylvania State University, University Park,
PA 16802, USA

The impact of international trade on the environment has been a contentious issue since the early 1990s. The debates over the North American Free Trade Agreement (NAFTA) and the Uruguay Round trade negotiations, which led to the creation of the World Trade Organization (WTO), stimulated a great deal of economic research on the ways in which international trade might be environmentally harmful or beneficial. Interest in the effects of trade on the environment then subsided until 1999, when protests in Seattle over a new round of WTO negotiations moved the issue to the forefront once again.

In theory, the environmental impacts of trade liberalization would not be a cause for concern if socially optimal, flexible and internationally coordinated environmental policies were in place in every country worldwide. In such a world, because environmental policies were flexible, any worsening of environmental externalities in some country due to trade liberalization would be mitigated by a socially optimal increase in the stringency and/or scope of that country's environmental policies. Furthermore, any worsening of a global environmental externality such as biodiversity loss would be mitigated by an internationally coordinated and socially optimal increase in the stringency and/or scope of environmental policies worldwide.

Whether these conditions can be satisfied to a reasonable degree is questionable, particularly in agriculture. One of the messages of the preceding chapters of this book is that designing cost-effective environmental policies that can adequately address environmental problems associated with agricultural production is a hard problem. The non-point character of agricultural pollution places constraints on the options available to policy-makers. One cannot control non-point pollution in the way that one can control point sources of pollution, such as the flow of sewage out of a pipe or pollutants

© CAB *International* 2001. *Environmental Policies for Agricultural*
Pollution Control (eds J.S. Shortle and D.G. Abler)

from a smokestack. In contrast to point source pollution, the assignment of responsibility is also difficult or impossible. How much a particular farm is polluting, or even whether it is polluting at all, is usually very uncertain. Moreover, the feasibility, effectiveness and cost of pollution prevention and control technologies tend to vary significantly from one location to another. These considerations tend to hinder the design and implementation of cost-effective environmental policies for agriculture along the lines of the 'polluter pays' principle.

Because of the difficulties in designing environmental policies for agriculture, one cannot count on environmental policies to buffer fully any negative environmental impacts of trade liberalization. Furthermore, one cannot count on any environmental policies that may be in place to be flexible – that is, to adjust in a socially optimal manner to changes in environmental externalities created by trade liberalization. By the same token, trade liberalization could have positive environmental impacts in the agricultural sector that environmental policies, because of their intrinsic design difficulties, could not achieve in a cost-effective manner.

The literature on trade liberalization and the environment clearly indicates that there is no simple and universal answer to whether agricultural trade is good or bad for the environment (Abler and Shortle, 1998; Antle and Capalbo, 1998; Ervin, 1999). Instead, a large number of effects come into play, some positive and some negative. Bearing this in mind, the objective of this chapter is to decompose the ways in which trade may affect the environment in a single country and, in particular, the environment as it relates to agriculture.

It should be noted at the outset that most of the environmental externalities associated with agriculture are domestic (indeed often local) in scope. Agricultural pollution, for the most part, is not a transborder externality – with some exceptions, no one outside of a country is directly harmed by pollution from that country's agriculture. One can ask why one country should object if another country willingly chooses to degrade its own environment. Developed countries themselves made substantially different choices in the past than today; witness the fact that most major environmental legislation in the US, EU and other developed countries dates from the 1960s and 1970s. Efforts to use trade agreements to force other countries toward developed-country environmental standards can smack of 'eco-imperialism' or thinly disguised attempts at protectionism (Bhagwati, 1995). Be that as it may, it seems clear that trade and the environment will be an important issue for some time to come, and given this it is clearly better to proceed on the basis of good information than on the basis of misinformation or no information.

A Critique of Existing Decomposition Frameworks

The negotiations and debates over the Uruguay Round Agreement (URA) and the NAFTA in the late 1980s and early 1990s spawned considerable interest among policy-makers, environmental groups and economists over trade–environment interactions. Based on earlier work by John Miranowski, Gene Grossman, Alan Krueger and others (e.g. Miranowski *et al.*, 1991; Grossman and Krueger, 1992), the Organisation for Economic Cooperation and Development (OECD) initiated a series of work directed at decomposing the environmental effects of trade.

In a well-known report, OECD (1994a) distinguished between so-called product effects, scale effects and structural effects of trade on the environment. Product effects refer to trade in specific products that have particularly beneficial or harmful environmental impacts. For example, positive product effects may result from 'environmentally friendly' consumer goods or inputs into production, while negative product effects may result from trade in such things as hazardous wastes.

Scale effects refer to positive or negative environmental impacts caused by changes in the overall scale of economic activity. On the one hand, negative scale effects can arise because additional production and consumption can lead to environmental degradation. On the other hand, positive scale effects may arise if, as some evidence suggests, wealthier countries are more likely to adopt environmental protection policies. The structural effects category is not clearly defined by OECD (1994a) but, essentially, it appears to be a residual, encompassing all effects not classified as product or scale effects. It would include, among many other things, international changes in the location, intensity and mix of production and consumption activities.

While this framework is a useful initial effort, dividing the environmental effects of trade into only three broad categories causes a variety of quite distinct effects to be lumped into each category. The negative effects of an expansion in the total scale of economic activity operate through economic channels that are very different from the political channels through which positive scale effects operate. Depending on economic and political institutions, there could also be significant differences in the time horizons over which these positive and negative effects manifest themselves. For example, if political institutions were slow to respond, the negative scale effects could occur well before the positive scale effects. In addition, as will be seen when we lay out our own decomposition framework, many effects do not fall under the product or scale effects headings. In the OECD (1994a) framework, one would be forced to lump all of these effects together under the heading of structural effects. These effects turn out to be quite distinct from each other in the ways that they operate and in the degree to which they are environmentally beneficial or harmful.

A subsequent report by OECD (1994b) contained a revised framework that comes somewhat closer to describing the full range of potential environmental effects of trade and to putting distinct effects into distinct categories. This framework divides the potential effects of trade into five categories: product, scale, structural, technology and regulatory.

Product and scale effects in the revised OECD framework are defined in essentially the same way as in the initial framework and, as such, the limitations of the original framework apply here as well. Like the initial framework, product effects in the revised framework refer to trade in specific products that have particularly beneficial or harmful environmental impacts. In addition, like the initial framework, scale effects in the revised framework are associated with changes in the overall level of economic activity.

Unlike the initial OECD framework, structural effects are not defined as a residual in the revised framework. Instead, they are effects associated with changes in the pattern or mix of economic activity. While the revised definition is clearer, the structural effects category now overlaps to some extent with the product effects category. Changes in the mix of economic activity in one country (structural effects) could manifest themselves as, among other things, increased trade in environmentally beneficial or harmful products (product effects) by that country. The result is that product effects are double-counted.

Technology effects refer to changes in production processes for goods and services. Positive technology effects may occur if multinationals transfer 'clean' technologies from one country to another, or if firms in one country purchase inputs embodying cleaner technologies from some other country. Negative technology effects can occur if, for some reason, dirty technologies are transferred instead of clean ones. Regulatory effects refer to the legal and policy effects of a trade agreement on environmental regulations or standards. For example, a trade agreement may include environmental provisions. Alternatively, a trade agreement might conceivably prevent a government from enacting certain types of environmental regulations. Perhaps surprisingly, the regulatory effects category does not include changes in environmental regulations due to the effects of trade on aggregate income and, in turn, societal demands for environmental quality. In OECD's revised framework, these changes continue to be classified as scale effects.

An Alternative Decomposition Framework

The purpose of any decomposition framework is to break a complex problem into smaller, more manageable and more understandable pieces. In order for a framework to make sense, the pieces must be mutually exclusive and exhaustive; in other words, they should not overlap and should, when added up, give the whole picture. It is also a great aid in understanding the original problem if each piece consists of similar effects rather than dissimilar effects. In this way, even if the original problem is highly complex, each of the individual pieces can be understood and clearly interpreted. On both these grounds, the frameworks proposed by OECD (1994a,b) are lacking. In this section, we lay out an alternative decomposition framework designed to better meet these two criteria, and illustrate the framework within a two-sector economic model.

Our framework decomposes the environmental effects of trade within a country into six mutually exclusive and exhaustive categories: scale, output mix, input mix, externality, policy and technology. As will become apparent, while much can be said both in theory and in practice about scale, output mix and input mix effects, much less can be said about the other three effects. Paradoxically, these remaining effects, especially technology effects, could turn out to be the most important.

Scale and output mix effects

Scale effects are environmental effects arising from a change in the total scale of economic activity, holding constant the mix of goods produced and consumed. Unlike the OECD (1994a,b) frameworks, the effects of trade on aggregate income and, in turn, societal demands for environmental quality are not classified as scale effects in our framework. Instead, they are part of the policy effects category. The output mix effects category captures environmental impacts owing to changes in the mix of goods produced and consumed, holding constant the total scale of economic activity. Both scale and output mix effects hold constant any environmental impacts due directly to changes in the mix of inputs used in the production of each good (input mix effects). Scale and output mix effects also hold constant the impacts of environmental externalities or other externalities on production and consumption (externality effects), environmental policies and other public policies (policy effects) and the technologies used in production (technology effects).

For a better understanding of scale and output mix effects, it is helpful to consider a relatively simple model that can be illustrated diagrammatically. Consider a market economy containing two sectors, which for expositional purposes will be referred to as agriculture (subscript A) and non-agriculture (subscript N). Some factors of production can be used in either agriculture or non-agriculture, while there may be other factors specific to one of the two sectors. In anticipation of the discussion below, we assume that the government, if it chooses, can adopt environmental policies toward producers in either or both sectors. For the sake of simplicity, the government is not assumed to adopt any other policies toward producers. In addition, in anticipation of the discussion to follow, we assume that environmental externalities from production may impair production possibilities in either or both sectors.

Output supply in sector $i (i = A, N)$ is:

$$S_i = S_i(p, \overline{E}_A, \overline{E}_N, Z_A, Z_N, A_A, A_N), \tag{1}$$

where $p = p_A/p_N$ is the relative price of agricultural goods, $\overline{E}_i$ is the stock of environmental capital used in production in sector i, Z_i is a scalar or vector of environmental policies toward sector i, and A_i is a scalar or vector of technologies used in sector i. We can trace out the economy's production possibilities frontier (PPF) by varying p and plotting the resulting combinations of S_A and S_N.

We have $\partial S_A/\partial p \geq 0$, $\partial S_N/\partial p \leq 0$, $\partial S_i/\partial \bar{E}_i \geq 0$, $\partial S_i/\partial Z_i \leq 0$ and $\partial S_i/\partial A_i \geq 0$. The sign of $\partial S_i/\partial A_j$ ($j \neq i$) is in general ambiguous. It depends on whether technical change in one sector draws factors of production into that sector from the other sector, or whether it pushes them out. For similar reasons, the sign of $\partial S_i/\partial \bar{E}_j$ ($j \neq i$) is ambiguous. The effect of environmental policies in one sector on output in the other sector ($\partial S_i/\partial Z_j$, $j \neq i$) is also ambiguous in sign. On the one hand, environmental policies in one sector may reduce returns to factors of production in that sector, causing factors to move to the other sector. On the other hand, many environmental policies require firms to devote inputs to pollution abatement. These inputs are no longer available for production in either sector. The potential environmental effects of changes in $\bar{E}_i$, Z_i and A_i are taken up below in the sections on externality effects and policy effects. For now, we assume that these variables are fixed at their initial (superscript 0) values $\bar{E}_i^0$, Z_i^0 and A_i^0.

On the consumer side, there is an aggregate indirect utility function $u = u(p, Y, \bar{E})$, where $Y = pS_A + S_N$ is aggregate income in units of the non-agricultural good and $\bar{E}$ is the total stock of environmental capital. In this respect, the environment is a public good that positively affects utility. From Roy's identity, we can obtain domestic consumer demands for agricultural and non-agricultural goods:

$$D_i = D_i(p, Y, \bar{E}) \tag{2}$$

Both goods are assumed to be normal, so that $\partial D_A/\partial p \leq 0$ and $\partial D_i/\partial Y \geq 0$. The effect of a change in p on D_N is ambiguous, as are the effects of a change in $\bar{E}$ on D_A and D_N. The potential effects of a change in $\bar{E}$ are taken up below in the section on externality effects. For the moment, we assume that $\bar{E}$ is fixed at its initial value $\bar{E}^0$.

Initially, assume that there is no trade, so that $S_i^0 = D_i^0$. Once the economy is opened up to trade (superscript 1), assume that it becomes a net importer of agricultural goods ($S_A^1 < D_A^1$) and a net exporter of non-agricultural goods ($S_N^1 > D_N^1$). These directions of change in trade for agriculture and non-agriculture as a whole are consistent with projections of the effects of global trade liberalization on the EU, the US and Japan (e.g. Goldin *et al.*, 1993; Anderson and Strutt, 1996). They imply that the relative domestic price of agricultural goods, which is now equal to the world price (p^w), falls because of trade ($p^1 = p^w < p^0$). To keep things simple and focused, assume that the country is 'small' in a trade sense, so that p^w is exogenous.

Domestic production and consumption before and after trade are illustrated diagrammatically in Fig. 7.1. As a result of trade, domestic agricultural output falls from S_A^0 to S_A^1, while non-agricultural output rises from S_N^0 to S_N^1. In order to decompose these movements into scale and output mix effects, draw a ray from the origin in Fig. 7.1 that goes through the point $S^1 = (S_A^1, S_N^1)$. Along this ray, the mix of goods produced is the same as the post-trade

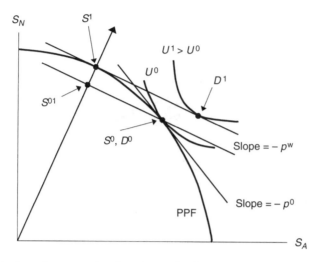

Fig. 7.1. Scale and output mix effects in production.

mix, in the sense that $S_N/S_A = S_N^1/S_A^1$; only the scale of production varies. At the same time, draw a line in Fig. 7.1 that intersects the point $S^0 = (S_A^0, S_N^0)$ and has a slope of $-p^w$. Along this line, the scale of production is the same as in the pre-trade situation, in the sense that $p^w S_A + S_N = p^w S_A^0 + S_N^0$; only the output mix varies. The intersection of this line and the ray from the origin define a new point $S^{01} = (S_A^{01}, S_N^{01})$. The movement from S^0 to S^{01} is then the output mix effect on domestic production, while the movement from S^{01} to S^1 is the scale effect. A similar decomposition can be carried out on the demand side and is illustrated in Fig. 7.2.

To analyse the environmental impacts of changes in scale and output mix, we need to describe how the stock of environmental capital is affected by production and/or consumption in each sector. For the sake of simplicity, the total environmental capital stock is assumed to be the sum of the stocks used as inputs in each sector:

$$E = E_A + E_N \tag{3}$$

Each sector's stock of environmental capital is negatively affected by production in that sector but is not affected by production in the other sector or by consumption. Each sector's stock is also affected by the mix of inputs used to produce any given level of output because some inputs may be more polluting than others. Let x_i be a scalar or vector characterizing the input mix (this will be made more precise in the section on input mix effects below). Environmental policies reduce pollution at any given level of production. Technical change might reduce or increase pollution at any given level of production, depending on the technologies employed. The result is that:

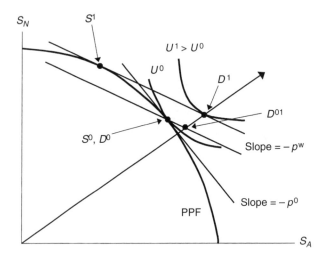

Fig. 7.2. Scale and output mix effects in consumption.

$$E_i = E_i(S_i, x_i, Z_i, A_i),\tag{4}$$

with $\partial E_i/\partial S_i \leq 0$ and $\partial E_i/\partial Z_i \geq 0$. Note that $\bar{E}_i$ and $\bar{E}$ are used as arguments in the supply and demand functions above, but E_i and E are used here. In the section below on externality effects, we will bring them together by setting $\bar{E}_i = E_i$ and $\bar{E} = E$, thereby allowing $\bar{E}_i$ and $\bar{E}$ to vary. For the time being, however, $\bar{E}_i$ and $\bar{E}$ are held fixed even though E_i and E vary. For the time being, x_i is also held fixed at its initial (superscript 0) value x_i^0. We will permit x_i to vary in equation (4) in the input mix effects section below.

The environmental impacts of changes in the mix of outputs depend on the relative impacts of changes in production in each sector on the total stock of environmental capital. Trade causes agricultural production to decrease and non-agricultural production to increase. If agricultural production is less polluting at the margin than non-agricultural production (in a sense to be made clear below), the output mix effect on the environment is negative. On the other hand, if agricultural production is more polluting at the margin, the output mix effect is positive. The environmental impacts of changes in the scale of production are always negative, because the scale effect works to increase both agricultural and non-agricultural output.

Starting at some point along the PPF, consider the trade-offs that we face between agricultural and non-agricultural production if the total stock of environmental capital is held constant in the equation $E = E_A + E_N$. The slope of this 'environment-constant' trade-off curve is:

$$(dS_N/dS_A)|_E = -(\partial E_A/\partial S_A)/(\partial E_N/\partial S_N)\tag{5}$$

The corresponding slope for the PPF is:

$$(dS_N/dS_A)|_{PPF} = (\partial S_N/\partial p)/(\partial S_A/\partial p) \qquad (6)$$

If the slope of the environment-constant trade-off curve is less in absolute value than the slope of the PPF, then a small upward movement along the PPF must reduce E. If this holds at all points on the PPF between S^0 and S^1, then over the relevant range we can say that agricultural production is less polluting at the margin than non-agricultural production. Alternatively, if the slope of the environment-constant trade-off curve is greater in absolute value than the slope of the PPF at all points between S^0 and S^1, then over the relevant range agricultural production is more polluting at the margin.

The two possibilities are illustrated in Figs 7.3 and 7.4, respectively. In the case where agricultural production is less polluting at the margin (Fig. 7.3), the output mix effect on the environment is negative ($E^{01} < E^0$) because trade shifts the composition of products toward more polluting products. The scale effect leads to a further loss in environmental capital ($E^1 < E^{01}$). In the case where agricultural production is more polluting at the margin (Fig. 7.4), the output mix effect on the environment is positive. The scale effect is negative, but the sum of the scale and output mix effects is still positive ($E^1 > E^0$).

Whether agriculture is more polluting or less polluting at the margin than non-agriculture is an empirical question that is beyond our scope here. For developed countries, one might reasonably argue that agriculture is less polluting in total than non-agriculture. Non-agriculture includes industry, and industry in developed countries is responsible for a wide array of water pollution, anthropogenic air pollution, greenhouse gas emissions, hazardous

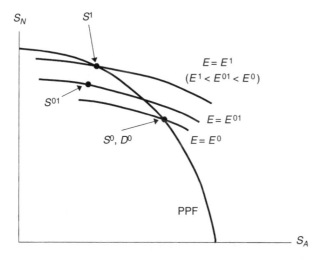

Fig. 7.3. Scale and output mix environmental effects (agriculture less polluting at the margin).

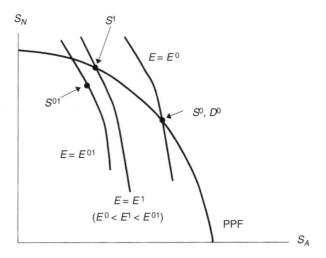

Fig. 7.4. Scale and output mix environmental effects (agriculture more polluting at the margin).

wastes, solid wastes, etc. (OECD, 1999). On the other hand, non-agriculture also includes the services sector, which for the most part is not a major polluter. In any event, what is relevant is not total pollution but rather changes in pollution from each sector in response to additional trade. In this regard, empirical evidence is lacking.

Simulation analyses of trade liberalization almost uniformly show very small impacts on the total scale of economic activity. For example, Nguyen *et al.* (1993) estimated that the Uruguay Round Agreement would lead to about a 1% increase in national income in Australia, Canada, New Zealand and the USA, and about a 2% increase in the EU and Japan. For the world as a whole, their estimated increase in total income was about 1%. Simulations by Goldin *et al.* (1993) indicated that full global trade liberalization would lead to about a 3% increase in aggregate income in the EU and Japan, and no significant change in Eastern Europe, Canada or the USA. Schott (1994) obtained similar findings.

Simulation analyses also tend to show only moderate effects of trade liberalization on output mix in most countries, at least at the aggregate level of agriculture and non-agriculture (e.g. Goldin *et al.*, 1993; Anderson and Strutt, 1996). Among developed countries, the major exception to this statement would be Japan. However, the story is very different in most countries if one looks at specific commodities that are highly protected from imports. For example, in a computable general equilibrium (CGE) analysis for Mexico, Beghin *et al.* (1997) found that trade liberalization would significantly reduce production of staple crops such as maize, beans and sorghum, and that this would be accompanied by a significant reduction in water and soil effluents from staple crops. The story might also be very different if one looked beyond changes in production at a national level to more geographically disaggregated levels that may be more relevant from an environmental perspective (Antle and Capalbo, 1998; Ervin, 1999). Small percentage changes in production at a national level could potentially mask significant percentage changes (positive or negative) in production in environmentally sensitive areas of a country.

Input mix effects

Input mix effects are environmental effects arising from changes in the mix of inputs used to produce any given level of output. These effects arise because some inputs may be more polluting than others. Input mix effects are derived holding constant scale and output mix effects, the impacts of environmental externalities or other externalities on production and consumption (externality effects), environmental policies and other public policies (policy effects) and the technologies used in production (technology effects). In terms of equation (4), input mix effects refer to the effects of trade liberalization on E_i due to changes in x_i, holding S_i, Z_i and A_i constant.

To simplify matters, suppose there are two inputs: one that is polluting and another that is not polluting. For heuristic reasons, refer to the polluting input as physical capital and the non-polluting input as human capital. Let K_i be the amount of physical capital in sector i, let H_i be the amount of human capital, and let $x_i = K_i/H_i$ represent the use of the polluting input relative to the non-polluting input, so that $\partial E_i/\partial x_i < 0$.

If physical capital and human capital can both move freely between the two sectors, agriculture and non-agriculture, and if markets for these two inputs are free from distortions, then prices for both inputs will be equal across the two sectors. Assuming these conditions hold, let ω be the economy-wide ratio of the price of human capital to the price of physical capital. Also assume for expositional purposes that production in each sector is characterized by constant returns to scale in physical and human capital. Then, following Mundlak (1970), it is possible to obtain a solution for ω in terms of p and other variables:

$$\omega = \omega(p, \bar{E}A, \bar{E}N, Z_A, Z_N, A_A, A_N) \tag{7}$$

It is also possible under these conditions to obtain a solution for x_i in terms of ω and other variables:

$$x_i = K_i/H_i = x_i(\omega, \bar{E}i, Z_i, A_i), \tag{8}$$

with $\partial x_i/\partial \omega > 0$. One would expect stricter environmental policies to reduce relative use of the polluting input ($\partial x_i/\partial Z_i < 0$), while the effect of an increase in $\bar{E}i$ or A_i on x_i is uncertain in sign. For the case in which there are factor market distortions, see Mundlak (1970).

The effect of a change in p on ω in equation (7) is governed by Stolper-Samuelson considerations. If agriculture is intensive in physical capital relative to non-agriculture ($x_A > x_N$), then an increase in agriculture's relative price will benefit physical capital at the expense of human capital ($\partial\omega/\partial p < 0$). On the other hand, if agriculture is less intensive in physical capital than non-agriculture ($x_A < x_N$), then an increase in agriculture's relative price will have the opposite effect ($\partial\omega/\partial p > 0$). Statistics for developed countries and many developing countries indicate that agriculture is intensive in physical capital relative to non-agriculture (UN Statistical Office, various years). Agriculture is generally not as capital intensive as industry, but it is significantly more capital intensive in most countries than the services sector, which comprises the vast majority of non-agriculture in most countries.

One would expect stricter environmental policies (an increase in Z_A or Z_N) to reduce the demand for the polluting input (physical capital) relative to the non-polluting input (human capital), leading to an increase in the relative price of the non-polluting input ($\partial\omega/\partial Z_i > 0$). The effects of changes in environmental capital stocks or technologies on ω are uncertain in sign.

Figure 7.5 illustrates input mix effects for each sector in the case where trade liberalization leads to a decline in the relative domestic price of agricultural goods (p falls) and where agriculture is relatively intensive in physical capital, so that ω increases in response to the fall in p. In Fig. 7.5, ω increases from its pre-trade liberalization value ω^0 to its post-trade liberalization value ω^1. In this case, if we hold domestic production in the ith sector (S_i) constant at its initial level, S_i^0, trade liberalization leads to a movement along an isoquant from (H_i^0, K_i^0) to (H_i^*, K_i^*). As K_i increases from K_i^0 to K_i^*, environmental capital in the ith sector falls from E_i^0 to E_i^*. Consequently, input mix effects are environmentally harmful in this case. Figure 7.5 would also illustrate input mix effects in the case where p increased and agriculture was less intensive in physical capital relative to non-agriculture, because in this case ω would still rise in response to trade liberalization.

On the other hand, input mix effects would be environmentally beneficial in the case where p decreased and agriculture was relatively intensive in physical capital. In such a case ω would fall, causing x_i to fall in both sectors. Input mix effects would also be environmentally beneficial in the case where p increased and agriculture was less intensive in physical capital than non-agriculture, because ω would also fall in such a case.

The picture becomes more complicated when we combine input mix effects with scale and output mix effects. In the case where p falls in response to trade liberalization, agricultural production decreases and non-agricultural production increases. One would expect the increase in non-agricultural production to lead to an increase in the derived demand for physical capital in non-agriculture, causing K_N to increase even more than it would based on the input mix effect alone. As this happened, E_N would fall even more than the fall from E_N^0 to E_N^* in Fig. 7.5. In agriculture, one would expect the decrease in agricultural production to lead to a decrease in the derived demand for physi-

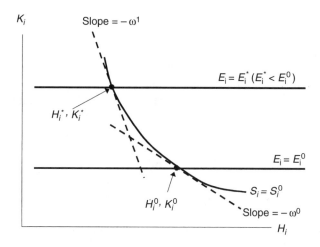

Fig. 7.5. Input mix environmental effects (agriculture intensive in physical capital relative to non-agriculture).

cal capital in agriculture. Whether or not this decrease in demand would overwhelm the increase in K_A due to the input mix effect is unclear. However, if we assumed that the economy's total supply of physical capital was fixed, K_A would have to fall because the increase in K_N would have to come out of agriculture.

Empirical studies of trade liberalization have generally not considered input mix effects in isolation. Instead, these studies have estimated effects of trade liberalization on environmental externalities in agriculture and other sectors that are implicitly a combination of input mix effects, output mix effects and scale effects. Some studies have found that these three effects in combination could be significant. For example, Abler and Shortle (1992) found that trade liberalization involving a complete elimination of price support programmes for major grains in the US and EU would significantly reduce fertilizer and pesticide use in EU agriculture but could work to increase their use in US agriculture. Using a CGE model for Sri Lanka, Bandara and Coxhead (1999) found that trade liberalization would shift agricultural land into the production of tea, which is not as erosive as most other crops in Sri Lanka. On the other hand, other studies have found that scale, output mix and input mix effects when taken together would on the whole be relatively small. For example, CGE analyses by Abler *et al.* (1999) for Costa Rica and Strutt and Anderson (1999) for Indonesia, using a wide range of environmental indicators, found that trade liberalization generally led to small changes in these indicators.

Externality effects

Externality effects refer to the feedback effects on production and/or consumption of environmental externalities or other externalities caused by production and/or consumption. In the two-sector model above, production in each sector generates environmental externalities and, in turn, negative feedback effects on production in that sector. These feedback effects were set aside in the discussion of scale, output mix and input mix effects by holding $\overline{E}_i$ and $\overline{E}$ constant. It is now time to bring these feedback effects into the picture by allowing $\overline{E}_i$ and $\overline{E}$ to vary, with $\overline{E}_i = E_i$ and $\overline{E} = E$.

Due to the nature of the externalities in the model, the method outlined above to measure scale and output mix effects does not change once externality effects are brought into the picture. However, starting from any point such as S^0 in Fig. 7.1, the shape of the PPF changes and, as a result, the scale, output mix and input mix effects change. There are a large number of possibilities. Suppose, to help to illustrate this point, that externalities are stronger at the margin at S^0 in non-agricultural production than in agricultural production. This causes the slope of the PPF at S^0 to be smaller in absolute value than it would be if $\overline{E}_A$ and $\overline{E}_N$ were held constant, because externalities limit what we can gain in non-agricultural output by giving up agricultural output. Moreover, if $\partial^2 E_i / \partial S_i^2 < 0$, the gap between the two slopes must grow as the economy moves toward more S_N and less S_A (the opposite would occur if the economy moved in the other direction). The result, as shown in Fig. 7.6, is that post-trade agricultural output is larger (point $(S^1)^*$) than it would be if $\overline{E}_A$ and $\overline{E}_N$ were held constant (S^1), while the opposite holds for non-agricultural output.

Changes in domestic consumption have no environmental implications in the model here because there are no externalities from consumption. In any event, the method diagrammed in Fig. 7.2 to decompose scale and output mix effects in consumption still holds once externality effects are brought into the picture. However, externality effects would cause the post-trade consumption point (D^1) to differ from the corresponding point in the absence of externality effects for two reasons. Firstly, externality effects will in general lead to a change in the post-trade production point (S^1) and thus post-trade aggregate income. Secondly, the total stock of environmental capital enters the utility function as a public good. The effects of a change in E on D_A and D_N are uncertain.

A richer model would permit many other types of externalities, both environmental and non-environmental. Consumers might be significant sources of environmental externalities in some cases – for example, because of household wastes or automobile emissions. These externalities tend to be substantial in developing countries, so that any study of a developing country would need to take them into account (World Bank, 1992; World Resources Institute, 1996). Alternatively, there might be externality effects in production that spill over from one sector to another. As Baumol and Oates (1988)

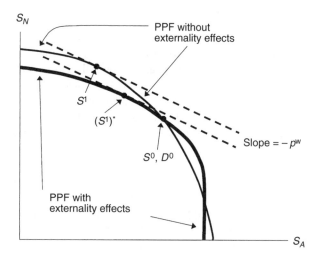

Fig. 7.6. Externality effects and the production possibilities frontier.

discussed, these types of intersectoral spillovers, if they are sufficiently strong, can lead to non-convexities in the production possibility set that can cause the usual efficiency properties of market equilibria to break down.

Policy effects

Policy effects refer to changes in environmental policies and other public policies induced by trade. Our policy effects category encompasses the OECD (1994b) regulatory effects category, as well as changes in environmental policies that the OECD (1994a,b) puts in its scale effects category. In analysing policy effects, we allow production, consumption and environmental capital to vary in response to environmental policies along the lines modelled above. However, technologies (the A_i) are still treated as constant and equal to their initial values, A_i^0.

Within the confines of our two-sector model, consider the social welfare-maximizing choices for the environmental policies and how those choices respond to trade. Beginning with the aggregate indirect utility function $u = u(p, Y, \bar{E})$, the marginal social benefit of some environmental policy Z_i is the gain in utility due to an increase in the total environmental capital stock, or:

$$MSB_i = (\partial u/\partial \bar{E})(dE/dZ_i) \tag{9}$$

The marginal social cost is the loss in utility due to a decrease in aggregate income and any change in the relative price of agricultural goods, or:

$$MSC_i = -(\partial u/\partial p)(dp/dZ_i) - (\partial u/\partial Y)(dY/dZ_i) \tag{10}$$

In the post-trade situation, $dp/dZ_i = 0$ because p is exogenous and equal to p^w. Attaining the social optimum requires that $MSB_i = MSC_i$ for all environmental policies adopted (some might not be adopted at all because marginal social costs exceed marginal social benefits even at a minimal level of adoption).

Marginal social cost might well be negative at low levels of Z_i because environmental policy, if properly designed, can correct for harmful externalities. Beyond a certain point, however, losses in output due to costs imposed by pollution abatement and pollution prevention requirements outweigh gains from correction of externalities.

The impacts of trade on the marginal social benefit and marginal social cost curves are, in general, uncertain. Consequently, policy effects on the environment are also uncertain. The change in Z_i depends on how each of the components of the marginal social benefit and marginal social cost equations above responds to the economic changes caused by trade. The change in the environmental capital stock depends on whether changes in the Z_i reinforce or offset the change in E that occurs with environmental policies held fixed.

If there is diminishing marginal utility to income ($\partial^2 u/\partial Y^2 < 0$), the increase in aggregate income caused by trade will, by itself, push down the marginal social cost curve for every environmental policy. Beyond this, changes in the Z_i depend in part on which sector is more polluting at the margin (see the discussion above on scale and output mix effects). Suppose that non-agriculture is more polluting at the margin, so that E declines in the absence of any offsetting changes in environmental policies (see Fig. 7.3). If there is diminishing marginal utility to environmental capital ($\partial^2 u/\partial \bar{E}^2 < 0$), the reduction in E will, by itself, push out the marginal social benefit curve for every environmental policy. On the other hand, suppose that agriculture is more polluting at the margin, so that E increases in the absence of any changes in environmental policies (see Fig. 7.4). If there is diminishing marginal utility to environmental capital, the increase in E will, by itself, push out the marginal social benefit curve for every environmental policy.

Of course, real-world environmental policies are not determined solely by social welfare but rather are based in part, sometimes in large part, on political considerations. Economists and political scientists have not yet come up with any widely accepted model of public choice. For the time being, the best approach for the trade practitioner is probably an empirical one.

In this regard, several studies in recent years have estimated the relationship between per capita income and a variety of indicators of environmental degradation and environmental quality. These studies are relevant to the trade–environment debate to the degree that trade raises per capita income. Some of these studies (e.g. Antle and Heidebrink, 1995; Grossman, 1995; Lucas, 1996; Hilton and Levinson, 1998) suggest an 'inverted-U' relationship between per capita income and several types of environmental degradation, a relationship widely known as the environmental Kuznets curve. Other studies do not find this inverted-U relationship (e.g. Vincent, 1997; Koop and Tole, 1999) or find that intervening variables such as literacy and political rights play an important role in the relationship (Torras and Boyce, 1998). Moreover, for those types of environmental degradation that do appear to follow the environmental Kuznets curve, the level at which more income turns from environmental harm to improvement varies substantially from one type of degradation to another.

Environmental policies can be affected directly by trade liberalization if environmental side agreements are included as part of a trade agreement. The environmental side agreement to NAFTA is a well-known example. Environmental policies can also be affected directly by trade liberalization if the provisions of a trade agreement constrain the operation of domestic environmental legislation (Trachtman, 1999). For instance, trade agreements may limit domestic laws that restrict imports of environmentally sensitive products (e.g. tuna caught in nets that are not dolphin-safe) or laws that subsidize environmentally beneficial products or production practices. Whether subsidies for environmentally beneficial agricultural production practices should fall into the 'Green Box' or 'Blue Box' of permitted policies or the 'Amber Box' of policies to be curtailed has been a source of controversy in the WTO negotiations.

One concern sometimes expressed about trade liberalization is that it could encourage the migration of 'dirty' industries from developed countries to developing countries in order to escape stricter environmental regulations in developed countries, a process sometimes called the pollution haven hypothesis or the 'race to the bottom'. However, empirical studies have found very little support for the pollution haven hypothesis (Low, 1992; Eskeland and Harrison, 1997; Xu, 1999). Firms base international location decisions on a wide array of factors and environmental policies typically play a very minor role in the decision-making process.

In general, the policy effects category includes not only changes in environmental policies but also changes in a whole host of other public policies. While environmental policies might be the most important for our purposes here, changes in other policies could also have environmental impacts. For example, trade liberalization could increase or decrease a country's total government revenue. A reduction in import tariffs would tend to decrease government revenue, while a reduction in export subsidies and an increase in total tax collections (due to an increase in national income) would tend to increase government revenue. If the net result of all these changes were a decrease in government revenue, the result might be offsetting increases in other taxes or reductions in government expenditures on certain programmes. If the net result was an increase in revenue, that might pave the way for reductions in other taxes or increases in government expenditures. Depending on the taxes and expenditures affected, there could (at least in principle) be significant environmental impacts.

Technology effects

Technology effects refer to impacts of trade on the environment caused by the development and adoption of new products, new production processes or new pollution abatement technologies. New products or new production processes might be entirely new or might be ones already in use in other countries. New products might be sold directly to consumers or might be used as capital equipment or intermediate inputs in the production of other goods. In analysing technology effects, we allow production, consumption, environmental capital and public policies to vary in response to technology along the lines modelled above. Our technology effects category is similar to the OECD's (1994b) category of the same name, but somewhat broader in the kinds of changes in technology that we consider.

Trade can lead to changes in technology for at least four reasons. Firstly, trade can lead to the international diffusion of technologies, potentially including so-called 'environmentally friendly' technologies (Grossman and Helpman, 1991; Low, 1992). Trade enables a country to purchase a larger variety of capital equipment and intermediate inputs embodying technologies not previously available. Trade also provides channels of communication through which producers can learn about, and then copy or adapt, technologies used in other countries. Coe and Helpman (1995) and Coe *et al.* (1997) found that spillovers in research and development (R&D) between countries are significant, and that they are stronger the more open an economy is to trade. Ben-David (1993) found that trade openness has increased the speed at which EU countries have converged toward each other in terms of per capita income. Using a CGE model for Costa Rica with a series of environmental indicators, Abler *et al.* (1999) found that international diffusion of technologies due to trade liberalization could lead to changes in shifts in production and input usage across sectors that worsen some environmental externalities and improve others.

Secondly, by enlarging potential market sizes, trade can enable firms to better exploit scale economies in R&D. Many innovations are characterized by substantial up-front R&D costs, followed by production costs that are small or trivial in comparison with R&D costs. In agriculture, biotechnology tends to follow this pattern.

Thirdly, by changing relative output prices, trade alters the incentives to do research in one sector versus another because the rate of return to output-increasing R&D depends positively on output prices. For example, if trade lowers relative agricultural product prices, then the incentives to invest in agricultural research fall relative to the incentives to invest in non-agricultural research. In a simulation analysis of the effects of elimination of government programmes for maize in the US, Abler (1996) found this type of effect at work for private sector R&D on maize.

Fourthly, changes in relative output prices due to trade can have Stolper-Samuelson effects on factor prices. A decline in the relative output price in a sector tends to decrease the prices of factors in which that sector is relatively intensive and to increase the prices of other factors. These changes in factor prices could in turn lead to changes in technologies along the lines predicted by the induced innovation model (Hayami and Ruttan, 1985). In this model, technologies tend to be developed and adopted that conserve on relatively expensive factors of production.

Agricultural production throughout the world is intensive in land relative to other sectors, except perhaps forestry in some cases. In the case of many countries that protect their agricultural sectors, this means that trade liberalization, by reducing relative agricultural product prices, would tend to reduce land rents. Simulations in Abler and Shortle (1992) suggest that the decline in land rents in the case of the EU could in fact be substantial. Historically, agricultural researchers in the public and private sectors have conserved on land by developing and improving inputs that increase yields – fertilizers, pesticides, hybrid seeds and irrigation technologies (Hayami and Ruttan, 1985). The future might or might not be the same. In any event, these are the very inputs that are associated with environmental degradation in many cases. Lower agricultural output prices could diminish the tendency toward yield-increasing technologies.

Within the confines of our two-sector model, technical change in one sector can have a direct effect on the stock of environmental capital in that sector, holding constant output and environmental policies (recall that $E_i = E_i(S_i, Z_i, A_i)$). It can also have indirect effects that operate through changes in S_i and Z_i. In addition, changes in technology in one sector can potentially affect output and environmental policies in the other sector. All of these effects would need to be taken into account in computing the change in the total stock of environmental capital ($E = E_A + E_N$).

Even in the case of environmentally friendly technical change ($\partial E_i / \partial A_i > 0$), the total effect (the direct effect plus indirect effects in both sectors) could be environmentally harmful. Like all types of technical change, environmentally friendly technical change will increase output in the sector benefiting from the change. This, by itself, is environmentally harmful because $\partial E_i / \partial S_i \leq 0$. Essentially, we have less pollution per unit of output but more output. Under many circumstances, the negative impact on E_i of the increase in S_i could actually outweigh the positive impact of the increase in A_i (Abler and Shortle, 1995).

The situation becomes even cloudier when we consider the possibility that environmental policies could change. By increasing aggregate income, technical change would tend to lead to stricter environmental policies for reasons discussed above in the policy effects section. Environmentally friendly technical change could lead regulatory authorities to adopt stricter environmental regulations for another reason as well. Environmentally friendly technologies reduce the costs to producers of stricter environmental regulations, thereby making stricter regulations more politically feasible. This process is sometimes referred to as 'ratcheting' in the environmental economics literature.

Multi-country Considerations

The framework for decomposing the environmental effects of trade outlined above pertains to a single country. An international assessment of trade and the environment would require one to apply this framework to several countries. Shifts in production between countries, especially production of pollution-intensive goods, could have important environmental implications. Environmental damages from any given level of production could differ from one country to another. The economic loss attached to any given level of damages could also differ. Although this issue is quite important, it is beyond our scope here. Copeland and Taylor (1994, 1995) discussed some of the impacts of trade on the international location of polluting industries and on global pollution in the case of transboundary pollutants. Abler and Pick (1993) provided an empirical illustration of the location issue in the case of fruit and vegetable production in the USA versus Mexico under NAFTA.

An international assessment of trade–environment interactions would also require consideration of environmental externalities from international transportation. Gabel (1994) discussed some of the impacts that trade liberalization can have on the volume and modes of international transportation and, in turn, the environment.

Conclusions

This chapter outlined a framework for dividing the environmental effects of trade on a country into scale, output mix, input mix, externality, policy and technology. It should be clear from inspection of this framework that a comprehensive environmental assessment of trade would be a difficult task. It would require analysing the environmental effects of trade on a sector-by-sector basis and then expressing these effects in monetary terms in order to make comparisons across sectors. Each of these effects can be quite complex, and different effects tend to move in different directions, even within a single sector. Paradoxically, the effects about which we know the least, and the effects most commonly left out of trade–environment analyses – externality, policy and technology – could be the most important. Technology, in particular, is the 'wild card' that has the potential to dominate all other effects.

Nevertheless, environmental data collection and environmental valuation in Western Europe, North America and some other parts of the world have progressed to the point where a comprehensive, reasonably accurate environmental assessment of trade is now feasible. As the world moves toward more liberalized trade, including liberalized agricultural trade, the challenge for researchers and for governments is to bring the available data and methodology to bear on this important problem.

References

Abler, D.G. (1996) Environmental policies and induced innovation: the case of agriculture. In: *Agricultural Markets: Mechanisms, Failures and Regulations*. Elsevier, Amsterdam.

Abler, D.G. and Pick, D. (1993) NAFTA, agriculture, and the environment in Mexico. *American Journal of Agricultural Economics* 75:794–798.

Abler, D.G. and Shortle, J.S. (1989) Cross compliance and water quality protection. *Journal of Soil and Water Conservation* 44:453–454.

Abler, D.G. and Shortle, J.S. (1991a) The political economy of water quality protection from agricultural chemicals. *Northeast Journal of Agricultural and Resource Economics* 20:54–60.

Abler, D.G. and Shortle, J.S. (1991b) Innovation and environmental quality: the case of EC and US agriculture. In: Dietz, F.J., Ploeg, F. van der and Straaten, J. van der (eds) *Environmental Policy and the Economy*. North-Holland, Amsterdam.

Abler, D.G. and Shortle, J.S. (1992) Environmental and farm commodity policy linkages in the US and EC. *European Review of Agricultural Economics* 19:197–217.

Abler, D.G. and Shortle, J.S. (1995) Technology as an agricultural pollution control policy. *American Journal of Agricultural Economics* 77:20–32.

Abler, D.G. and Shortle, J.S. (1996) Environmental aspects of agricultural technology. In: Alston, J. and Pardey, P. (eds) *Global Agricultural Science Policy for the Twenty-first Century*. Conference Secretariat on Global Agricultural Science Policy for the Twenty-First Century, Melbourne, Australia.

Abler, D.G. and Shortle, J.S. (1998) Decomposing the effects of trade on the environment. In: *Agriculture, Trade and the Environment*. Edward Elgar, Cheltenham, UK.

Abler, D.G., Rodríguez, A.G. and Shortle, J.S. (1999) Trade liberalization and the environment in Costa Rica. *Environment and Development Economics* 4:357–373.

Abrahams, N. and Shortle, J.S. (2000) Uncertainty and the choice between compliance measures and instruments in nitrate nonpoint pollution control. Working Paper, Department of Agricultural Economics and Rural Sociology, Pennsylvania State University, University Park, Pennsylvania.

Adamowicz, W., Louviere, J. and Williams, M. (1994) Combining stated and revealed preference methods for valuing environmental amenities. *Journal of Environmental Economics and Management* 26:271–292.

Adamowicz, W., Boxall, P., Williams, M. and Louviere, J. (1998) State preference approaches for measuring passive use values: choice experiments and contingent valuation. *American Journal of Agricultural Economics* 80:64–75.

Adams, R.M. and Crocker, T.D. (1991) Materials damages. In: Braden, J.B. and Kolstad, C.D. (eds) *Measuring the Demand for Environmental Quality*. North-Holland, Amsterdam.

Adler, R.W. (1994) Reauthorizing the Clean Water Act: looking to tangible values. *Water Resources Bulletin* 30(5):799–807.

Alexander, R.B., Smith, R.A. and Schwarz, G.E. (2000) Effects of stream channel size on the delivery of nitrogen to the Gulf of Mexico. *Nature* 403:758–761.

Alston, J.M. and Hurd, B.H. (1990) Some neglected social costs of government spending in farm programs. *American Journal of Agricultural Economics* 72(1):149–156.

Andersen, M.S. (1999) Governance by green taxes: implementing clean water policies in Europe 1970–1990. *Environmental Economics and Policy Studies* 2:39–63.

Anderson, J.L., Bergsrud, F.G. and Ahles, T.M. (1995) Evaluation of the farmstead assessment system (FARM*A*SYST) in Minnesota. In: *Clean Water–Clean Environment–21st Century*, Vol. III, *Practices, Systems, and Adoption*. Proceedings of the conference Clean Water–Clean Environment–21st Century, Working Group on Water Quality, USDA, March 5–8, Kansas City, Missouri, pp. 9–12.

Anderson, K. and Strutt. A. (1996) On measuring the environmental impacts of agricultural trade liberalization. In: *Agriculture, Trade and the Environment: Discovering and Measuring the Critical Linkages*. Westview, Boulder, Colorado.

Anderson, R. and Rockel, M. (1991) Economic valuation of wetlands. Discussion Paper No. 65, American Petroleum Institute, Washington, DC.

Anderson, R.C., Lohof, A.Q. and Carlin, A. (1997) *The United States Experience with Economic Incentives in Environmental Pollution Control Policy*. US Environmental Protection Agency, Washington, DC.

Anonymous (1994) *Cryptosporidium* and public health. *Health and Environment Digest* 8(8):61–63.

Antle, J.M. (1984) The structure of US agricultural technology, 1910–78. *American Journal of Agricultural Economics* 66:414–421.

Antle, J.M. and Capalbo, S.M. (1998) Quantifying agriculture–environment tradeoffs to assess environmental impacts of domestic and trade policies. In: *Agriculture, Trade and the Environment*. Edward Elgar, Cheltenham, UK.

Antle, J.M. and Heidebrink, G. (1995) Environment and development: theory and international evidence. *Economic Development and Cultural Change* 43:603–625.

Antle, J.M. and Just, R.E. (1991) Effects of commodity program structure on resource use and the environment. In: Just, R.E. and Bockstael, N. (eds) *Commodity and Resource Policies in Agricultural Systems*. Springer-Verlag, Berlin.

Antle, J.M. and Just, R.E. (1993) Conceptual and empirical foundations for agricultural environmental policy analysis. *Journal of Environmental Quality* 21:307–316.

Antle, J.M. and Pingali, P.L. (1994) Pesticides, productivity, and farmer health: a Philippine case study. *American Journal of Agricultural Economics* 76(3): 418–430.

Arnold, J.G., Williams, J.R., Srinivasan, R. and King, K.W. (1995) *SWAT: Soil and Water*

Assessment Tool. US Department of Agriculture, Agricultural Research Service and Texas A&M University College Station, Texas.

Arrow, K., Solow, R., Portney, P., Leamer, E., Radner, R. and Schuman, H. (1993) Report of the NOAA Panel on Contingent Valuation. Unpublished paper, Resources for the Future, Washington, DC.

Atkinson, A.B. and Stiglitz, J.E. (1980) *Lectures on Public Economics.* McGraw-Hill Book Co., New York.

Babcock, B.A., Lakshminarayan, P.G., Wu, J. and Zilberman, D. (1997) Targeting tools for the purchase of environmental amenities. *Land Economics* 73(3):325–339.

Bailey, G.W. and Swank, R.R. Jr (1983) Modeling agricultural nonpoint source pollution: a research perspective. In: Schaller, F.W. and Bailey, G.W. (eds) *Agricultural Management and Water Quality.* Iowa State University Press, Ames, Iowa.

Baldock, D. and Mitchell, K. (1995) *Cross-compliance Within the Common Agricultural Policy.* Institute for European Environmental Policy, London.

Bandara, J.S. and Coxhead, I. (1999) Can trade liberalization have environmental benefits in developing country agriculture? A Sri Lankan case study. *Journal of Policy Modeling* 21:349–374.

Barbash, J.E. and Resek, E.A. (1996) *Pesticides in Ground Water: Distribution, Trends, and Governing Factors.* Pesticides in the Hydrologic System Series, Vol. 2. Ann Arbor Press, Chelsea, Michigan. 590 pp.

Bartfeld, E. (1993) Point-nonpoint source trading: looking beyond potential cost savings. *Environmental Law* 23(43):43–106.

Bartik, T.J. (1988a) Evaluating the benefits of non-marginal reductions in pollution using information on defensive expenditures. *Journal of Environmental Economics and Management* 15:111–127.

Bartik, T.J. (1988b) Measuring the benefits of amenity improvements in hedonic price models. *Land Economics* 64:172–183.

Batie, S. (1994) Designing a successful voluntary green support program: what do we know? In: Lynch, S. (ed.) *Designing Green Support Programs.* Henry Wallace Institute for Alternative Agriculture, Greenbelt, Maryland.

Batie, S.S., Cox, W.E. and Diebel, P.L. (1989) *Managing Agricultural Contamination of Ground Water State Strategies.* National Governors' Association, Washington, DC.

Batie, S. and Sappington, A. (1986) Cross-compliance as a soil conservation strategy: a case study. *American Journal of Agricultural Economics* 68(4):880–885.

Baumol, W. and Oates, W. (1988) *The Theory of Environmental Policy.* Cambridge University Press, Cambridge, UK.

Beach, E.D. and Carlson, G.A. (1993) A hedonic analysis of herbicides: do user safety and water quality matter? *American Journal of Agricultural Economics* 75:612–623.

Beavis, M. and Walker, M. (1983) Achieving environmental standards with stochastic discharges. *Journal of Environmental Economics and Management* 10:103–111.

Becker, H. (1990a) *Influencing Environmental Quality by Induced Innovations in Agricultural Production Systems.* Federal Agricultural Research Centre, Braunschweig, Germany.

Becker, H. (1990b) *Economic Impacts of Nitrate and Pesticide Regulation on Productivity in Agricultural Production Systems.* Federal Agricultural Research Centre, Braunschweig, Germany.

Beghin, J., Dessus, S., Roland-Holst, D. and Mensbrugghe, D. van der (1997) The trade and environment nexus in Mexican agriculture. A general equilibrium analysis. *Agricultural Economics* 17:115–131.

Bellinder, R.R., Gummesson, G. and Karlsson, C. (1993) Percentage-driven government mandates for pesticide reduction: the Swedish model. *Weed Technology* 8:350–359.

Ben-David, D. (1993) Equalizing exchange: trade liberalization and income convergence. *Quarterly Journal of Economics* 108:653–679.

Bernardo, D.J., Mapp, H.P., Sabbagh, G.J., Geleta, S., Watkins, K.B., Elliott, R.L. and Stone, J.F. (1993) Economic and environmental impacts of water quality protection policies. 2. Application to the Central High Plains. *Water Resources Research* 29(9):3081–3091.

Bhagwati, J. (1995) *Environment and Labour Standards, and Competition Policy: the New Issues Before the World Trading System*. Bureau of Economic Studies Occasional Paper No. 15, Macalester College, St Paul, Minnesota.

Bishop, R. (1994) A local agency's approach to solving the difficult problem of nitrate in the groundwater. *Journal of Soil and Water Conservation* 49(2 ss):82–84.

Boadway, R.W. and Bruce, N. (1984) *Welfare Economics*. Basil Blackwell, Oxford.

Bockstael, N., Hanemann, M. and Kling, C. (1987) Estimating the value of water quality improvements in a recreational demand framework. *Water Resources Research* 23:951–960.

Bockstael, N., Strand, I., McConnell, K. and Arsanjani, F. (1990) Sample selection bias in the estimation of recreation demand functions: an application to sportfishing. *Land Economics* 66:40–49.

Bockstael, N., McConnell, K. and Strand, I. (1991) Recreation. In: Braden, J. and Kolstad, C. (eds) *Measuring the Demand for Environmental Quality*. Elsevier, Amsterdam.

Bohm, P. and Russell, C. (1985) Comparative analysis of alternative policy instruments. In: Kneese, A. and Sweeny, J. (eds) *Handbook of Natural Resource and Energy Economics*, Vol. 1. Elsevier Science Publishers, Amsterdam.

Bord. R.J., Fisher, A., O'Connor, R.E. and Wheeler, W.J. (1993) *Fresh Water Quality, Quantity, and Availability: an American Policy Perception*. Environmental Resources Research Institute, Pennsylvania State University, University Park, Pennsylvania.

Bosch, D., Cook, Z. and Fuglie, K. (1995) Voluntary versus mandatory agricultural policies to protect water quality: adoption of nitrogen testing in Nebraska. *Review of Agricultural Economics* 17(2):13–24.

Bouzaher, A., Braden, J.B. and Johnson, G. (1990) A dynamic programming approach to a class of nonpoint pollution problems. *Management Science* 36:1–15.

Braden J.B. and Segerson, K. (1993) Information problems in the design of nonpoint pollution. In: Russell, C.S. and Shogren, J.F. (eds) *Theory, Modeling and Experience in the Management of Nonpoint-Source Pollution*. Kluwer Academic Publishers, Dordrecht, The Netherlands.

Braden, J.B., Bozaher, A., Johnson, G. and Miltz, D. (1989) Optimal spatial management of agricultural pollution. *American Journal of Agricultural Economics* 71:404–413.

Braden, J.B., Larson, R. and Herricks, E. (1991) Impact targets vs discharge standards in agricultural pollution management. *American Journal of Agricultural Economics* 73:388–397.

Brannlund, R. and Kristrom, B. (1999) *Economic Instruments in Environmental Policy in the Nordic Countries.* Arbetsrapport 277, Intitutionen for Skogsekonomi, SLU, Umea, Sweden.

Breembroek, J.A., Poppe, B.K.J. and Wossink, G.A.A. (1996) Environmental farm accounting: the case of the Dutch nutrients accounting system. *Agricultural Systems* 51(1):29–40.

Bricker, S.B., Clement, C.G., Pirhall, D.E., Orlando, S.P. and Farrow, D.R.G. (1999) *National Estuarine Eutrophication Assessment: Effects of Nutrient Enrichment in the Nation's Estuaries.* NOAA, Special Projects Office and the National Centers for Coastal Ocean Science, Silver Spring, Maryland. 71 pp.

Bromley, D. and Hodge, I. (1992) Private property rights and presumptive policy entitlements: reconsidering the premises of rural policy. *European Review of Agricultural Economics* 17:197–214.

Broussard, W. and Grossman, M. (1990) Legislation to abate pollution from manure: the Dutch approach. *North Carolina Journal of International Law and Commercial Regulation* 15:86–114.

Brouwer, F.M., Terluin, I.J. and Godeschalk, F.E. (1994) *Pesticides in the EC.* Agricultural Economics Research Institute (LEI-DLO), The Hague.

Browne, W.P. (1995) *Cultivating Congress.* University of Kansas, Lawrence, Kansas.

Browning, E.K. (1976) The marginal cost of public funds. *Journal of Political Economy* 84:283–298.

Bull, L. and Sandretto, C. (1995) The economics of agricultural tillage systems. In: *Farming for a Better Environment.* Soil and Water Conservation Society of America, Ankeny, Iowa, pp. 35–40.

Byström, O. and Bromley, D.W. (1998) Contracting for nonpoint-source pollution abatement. *Journal of Agricultural and Resource Economics* 23(1):39–54.

Cabe, R. and Herriges, J. (1992) The regulation of nonpoint-source pollution under imperfect and asymmetric information, *Journal of Environmental Economics and Management* 22:34–146.

Camacho, R. (1991) *Financial Cost-effectiveness of Point and Nonpoint Source Nutrient Reduction Technologies in the Chesapeake Bay Basin.* ICPRB Report, 91–8.

Camboni, S.M. and Napier, T.L. (1994) Socioeconomic and farm structure factors affecting frequency of use of tillage systems. Invited paper presented at the Agrarian Prospects III symposium, Prague, Czech Republic, September.

Cantor, K., Blair, A. and Zhan, S. (1988) Health effects of agrichemicals in ground-water: what do we know? In: *Agricultural Chemicals and Groundwater Protection: Emerging Management and Policy.* Freshwater Foundation, Navarre, Minnesota, pp. 27–42.

Carpentier, C.L., Bosch, D.J. and Batie, S.S. (1998) Using spatial information to reduce costs of controlling nonpoint source pollution. *Agricultural and Resource Economics Review* 27:72–84.

Carsel, R., Smith, C.N., Mulkey, L.A., Dean, J.D. and Jowise, P.P. (1984) *User's Manual for Pesticide Root Zone Model (PRZM): Release 1.* EPA-600/3-84-109, US Environmental Protection Agency, Athens, Georgia.

CAST (1996) *Integrated Animal Waste Management.* Task Force Report No. 128, November, Council for Agricultural Science and Technology, Ames, Iowa.

Caswell, M., Lichtenberg, E. and Zilberman, D. (1990) The effects of pricing policies on water conservation and drainage. *American Journal of Agricultural Economics* 72:883–890.

Caulkins, P.P., Bishop, R.C. and Bouwes, N.W. (1985) Omitted cross-price variable biases in the linear travel cost model: correcting common misperceptions. *Land Economics* 61:182–187.

CDC (1996) *Surveillance for Waterborne-Disease Outbreaks – United States, 1993–1994.* 45(SS-1), Centers for Disease Control and Prevention, Atlanta, Georgia.

Chambers, R.G. (1988) *Applied Production Analysis: a Dual Approach.* Cambridge University Press, Cambridge, UK.

Chesapeake Bay Program (1995) *The State of the Chesapeake Bay 1995.* Printed for the Chesapeake Bay Program by the US Environmental Protection Agency, Annapolis, Maryland.

Chesapeake Bay Program (1999) *The State of the Chesapeake Bay.* EPA-903-R99-013, US Environmental Protection Agency, Annapolis, Maryland, 62 pp.

Chesters, G. and Schierow, L. (1985) A primer on nonpoint pollution. *Journal of Soil and Water Conservation* 40:9–13.

Claassen, R. and Horan, R.D. (2001) Uniform and non-uniform second-best input taxes: the significance of market price effects on efficiency and equity. *Environmental and Resource Economics* (in press).

Coe, D.T. and Helpman, E. (1995) International R&D spillovers. *European Economic Review* 39:859–887.

Coe, D.T., Helpman, E. and Hoffmaister, A.W. (1997) North-south R&D spillovers. *Economic Journal* 107:134–149.

Conant, C.K., Duffy, M.D. and Holub, M.A. (1993) *Tradeoffs between Water Quality and Profitability in Iowa Agriculture.* University of Iowa, Iowa City, Iowa,

Cook, K., Hug, A., Hoffman, W., Taddese, A., Hinkle, M. and Williams, C. (1991) Center for Resource Economics and National Audubon Society. Statement before the Subcommittee on Environmental Protection, Committee on Environment and Public Works, US Senate, 17 July.

Cooper, C.M. (1993) Biological effects of agriculturally derived surface water pollutants on aquatic systems – a review. *Journal of Environmental Quality* 22:402–408.

Copeland, B.R. and Taylor, M.S. (1994) North–south trade and the environment. *Quarterly Journal of Economics* 109:755–787.

Copeland, B.R. and Taylor, M.S. (1995) Trade and transboundary pollution. *American Economic Review* 85:716–737.

Copeland, C. and Zinn, J.A. (1986) Agricultural nonpoint pollution policy: a federal perspective. Paper prepared for a colloquium on Agrichemical Management to Protect Water Quality, Washington, DC.

Courant, P.N. and Porter, R.C. (1981) Averting expenditure and the cost of pollution. *Journal of Environmental Economics and Management* 8:321–329.

Coyne, A. and Adamowicz, W. (1992) Modelling choice of site for hunting bighorn sheep. *Wildlife Society Bulletin* 20:26–33.

Cropper, M.L., Evans, W.N., Berardi, J.J., Duclas Soares, M.M. and Portney, P.R. (1992) The determinants of pesticide regulation: statistical analysis of EPA decision making. *Journal of Political Economy* 100:175–197.

Crowder, B.M. (1987) Economic costs of reservoir sedimentation: a regional approach to estimating cropland erosion damage. *Journal of Soil and Water Conservation* 42(3):194–197.

Crutchfield, S.R., Cooper, J.C. and Hellerstein, D. (1997) *Benefits of Safer Drinking Water: the Value of Nitrate Reduction.* AER-752, US Department of Agriculture, Economic Research Service, Washington, DC. June.

Dabbert, S., Dubgaard, A., Slangen, L. and Whitby, M. (1998) *The Economics of Landscape and Wildlife Conservation*. CAB International, Wallingford, UK.

Darwin, R.F. (1992) Natural resources and the Marshallian effects of input-reducing technological changes. *Journal of Environmental Economics and Management* 23:201–215.

Davies, J. and Mazurek, J. (1998) *Pollution Control in the United States: Evaluating the System*. Resources for the Future Inc., Washington, DC.

Desvousges, W.H., Johnson, F.R., Dunford, R.W., Boyle, K.J., Hudson, S.P. and Wilson, K.N. (1992) Measuring natural resource damages with contingent valuation: tests of validity and reliability. Paper presented at the Cambridge Economics, Inc. Symposium, *Contingent Valuation: a Critical Assessment*, Washington, DC, April.

DETR (1997) *Economic Instruments for Water Pollution*. Department of the Environment, Transport and the Regions, London.

Diamond, P.A., Hausman, J.A., Leonard, G.K. and Denning, M.A. (1992) Does contingent valuation measure preference? Experimental evidence. Paper presented at the Cambridge Economics, Inc. Symposium, *Contingent Valuation: a Critical Assessment*, Washington, DC, April.

Dicks, M.R. (1986) What will it cost farmers to comply with conservation compliance? *Agricultural Outlook*, No. 561, p. 25.

Dietz, F.J. and Hoogervorst, N.J.P. (1991) Toward a sustainable and efficient use of manure in agriculture: the Dutch case. *Environmental and Resource Economics* 1:313–332.

Dinham, B. (1993) *The Pesticide Hazard: a Global Health and Environmental Audit*. Zed Books, London.

Doering, O.C., Diaz-Hermelo, F., Howard, C., Heimlich, R., Hitzusen, F., Kazmierczak, R., Lee, J., Libby, L., Milon, W., Prato, A. and Ribaudo, M. (1999) *Evaluation of the Economic Costs and Benefits of Methods for Reducing Nutrient Lodas to the Gulf of Mexico: Topic 6 Report for the Integrated Assessment on Hypoxia in the Gulf of Mexico*. NOAA Coastal Ocean Program Decision Analysis Series No. 20. NOAA Coastal Ocean Program, Silver Spring, Maryland.

Donigian, A.S., Bicknell, B.R., Patwardhan, A.S., Linker, L.C. and Chang, C. (1994) *Chesapeake Bay Program Watershed Model Application to Calculate Bay Nutrient Loadings – Final Facts and Recommendations*. EPA-903-R-94-042, US Environmental Protection Agency, Chesapeake Bay Program Office, Annapolis, Maryland, 283 pp.

Dosi, C. and Moretto, M. (1993) NPS pollution, information asymmetry, and the choice of time profile for environmental fees. In: Russell, C.S. and Shogren, J.F. (eds) *Theory, Modeling and Experience in the Management of Nonpoint-Source Pollution*. Kluwer Academic Publishers, Dordrecht, The Netherlands.

Dosi, C. and Moretto, M. (1994) Nonpoint source externalities and polluter's site quality standards under incomplete information. In: Tomasi, T. and Dosi, C. (eds) *Nonpoint Source Pollution Regulation: Issues and Policy Analysis*. Kluwer Academic Publishers, Dordrecht, The Netherlands.

Downing, B.P. and White, L.J. (1986) Innovation in pollution control. *Journal of Environmental Economics and Management* 13:18–29.

Dubgaard, A. (1999) The Danish pesticide program: success or failure depending on indicator price. Paper to World Congress of Environmental and Resource Economists, Venice, June.

Dubgaard, A. (1994) The Danish environmental programs: an assessment of policy instruments and results. In: Napier, T.L., Camboni, S.M. and El-Swaify, S.A. (eds) *Adopting Conservation on the Farm*. Soil and Water Conservation Society, Ankeny, Iowa.

Duffield, J.W. and Patterson, D.A. (1991) Field testing existence values: an instream flow trust fund for Montana rivers. Paper presented at the annual meeting of the American Economic Association, New Orleans, January.

Eiswerth, M.E. (1993) Regulatory/economic instruments for agricultural pollution: accounting for input substitution. In: Russell, C.S. and Shogren, J.F. (eds) *Theory, Modeling and Experience in the Management of Nonpoint-Source Pollution*. Kluwer Academic Publishers, Dordrecht, The Netherlands.

ELI (1997) *Enforceable State Mechanisms for the Control of Nonpoint Source Water Pollution*. Environmental Law Institute, Washington, DC.

Elmore, T., Jaksch, J. and Downing, D. (1985) Point/nonpoint source trading programs for the Dillon Reservoir and planned extensions for other areas. In: *Perspectives on Nonpoint Source Pollution*. US Environmental Protection Agency, Washington, DC.

Engler, R. (1993) Lists of chemicals evaluated for carcinogenic potential. Memorandum, Office of Prevention, Pesticides, and Toxic Substances, US Environmental Protection Agency, 31 August, Washington, DC.

Englin, J. and Mendelsohn, R. (1991) A hedonic travel cost analysis for valuation of multiple components of site quality: the recreation value of forest management. *Journal of Environmental Economics and Management* 21:275–290.

Ervin, D.E. (1995) A new era of water quality management in agriculture: from best management practices to watershed-based whole farm approaches? *Water Resources Update* 101:18–28.

Ervin, D.E. (1999) Taking stock of methodologies for estimating the environmental effects of liberalized agricultural trade. Paper presented at OECD Workshop on Methodologies for Environmental Assessments of Trade Liberalization Agreements, http://www.oecd.org/ech/26–27oct/docs/3.pdf

Ervin, D.E. and Graffy, E.A. (1995) Technology for production and environmental quality: are we missing opportunities for complementarity? *Journal of Soil and Water Conservation* July–August: 352–353.

Ervin, D.E. and Graffy, E.A. (1996) Leaner environmental policies for agriculture. *Choices*, 1996, 4th quarter: 27–33.

Ervin, D.E. and Mill, J.W. (1985) Agricultural land markets and soil erosion: policy relevance and conceptual issues. *American Journal of Agricultural Economics* 67: 938–942.

Eskeland, G. and Harrison, A.E. (1997) Moving to greener pastures? Multinationals and the pollution haven hypothesis. http://www.worldbank.org/html/dec/Publications/Workpapers/WPS1700series/wps1744/wps1744.pdf, accessed March 2000.

Esseks, J.D. and Kraft, S.E. (1993) Opinions of conservation compliance held by producers subject to it. In: *Research and Development: Public and Private Investments Under Alternative Markets and Institutions*. Report for the American Farmland Trust, February, Washington, DC.

Faeth, P. (2000) *Fertile Ground: Nutrient Trading's Potential to Cost-effectively Improve Water Quality*. World Resources Institute, Washington, DC.

FAO (1996) Control of water pollution from agriculture – FAO Irrigation and Drainage Paper 55. Food and Agriculture Organization of the United Nations http://www.fao.org/docrep/W2598E/W2598E00.htm, accessed 22 November, 2000.

FAO (2000) *Crops and Drops: Making the Best Use of Land and Water*. Food and Agriculture Organization of the United Nations, Rome.

Fawcett, R.S., Christensen, B.R. and Tierney, D.P. (1994) The impact of conservation tillage on pesticide runoff into surface water: a review and analysis. *Journal of Soil and Water Conservation* 49:126–135.

Fawson, C. and Shumway, C.R. (1992) Endogenous regional agricultural production technologies. *Applied Economics* 24:1263–1273.

Feather, P., Hellerstein, D. and Hansen, L. (1999) *Economic Valuation of Environmental Benefits and the Targeting of Conservation Programs: The Case of the CRP*. Agricultural Economics Report 778, US Department of Agriculture, Economic Research Service, April.

Feather, P.M. and Amacher, G. (1993) Role of information in the adoption of best management practices for water quality improvement. *Agricultural Economics* 11:159–170.

Feather, P.M. and Cooper, J. (1995) *Voluntary Incentives for Reducing Agricultural Nonpoint Source Water Pollution*. US Department of Agriculture, Economic Research Service, Agriculture Information Bulletin No. 716. Government Printing Office, Washington, DC.

Fernandez-Cornejo, J. and Jans, S. (1999) *Pest Management in US Agriculture*. Agricultural Handbook No. 717. US Development of Agriculture, Economic Research Service. Government Printing Office, Washington, DC.

Fernandez-Cornejo, J., Jans, S. and Smith, M. (1998) Issues in the economics of pesticide use in agriculture: a review of the empirical evidence. *Review of Agricultural Economics* 20 (Autumn/Winter):462–488.

Fernandez-Cornejo, J. and McBride, W.D. (2000) *Genetically Engineered Crops for Pest Management in US Agriculture: Farm-Level Effects*. Agricultural Economics Report No. 786, US Department of Agriculture, Economic Research Service. Government Printing Office, Washington, DC.

Fisher, A. (1991) Risk communication challenges. *Risk Analysis* 11:173–179.

Fisher, A. and Raucher, R. (1984) Intrinsic benefit of improved water quality: conceptual and empirical perspectives. In: Smith, V.K. (ed.) *Advances in Applied Micro-Economics*. JAI Press Inc., Greenwich, Connecticut.

Flemming, R.A. and Adams, R.M. (1997) The importance of site-specific information in the design of policies to control pollution. *Journal of Environmental Economics and Management* 33:347–358.

Fontein, P.F., Thijssen, G.J., Magnus, J.R. and Dijk, J. (1994) On levies to reduce the nitrogen surplus: the case of the Dutch pig farm. *Environmental and Resource Economics* 4:445–478.

Fox, G., Weersink, A., Sarwar, G., Duff, S. and Deen, B. (1991) Comparative economics of alternative agricultural production systems: a review. *Northeastern Journal of Agricultural and Resource Economics* 20:124–142.

Franco, J., Schad, S. and Cady, C.W. (1994) California's experience with a voluntary approach to reducing nitrate contamination of groundwater: the Fertilizer Research and Education Program (FREP). *Journal of Soil and Water Conservation* 49(2 ss):76–81.

Frederikson, B. (1997) Legislation in response to the Nitrates Directive: aspects for some EU countries. In: Brouwer, F. and Kleinhanss, W. (eds) *The Implementation of Nitrate Policies in Europe*. Wissenschaftsverlag Vuak Kiel, Kiel, Germany.

Freeman, A.M. III (1979a) Hedonic prices, property values and measuring environmental benefits: a survey of the issues. *Scandinavian Journal of Economics* 1979:154–173.

Freeman, A.M. III (1979b) *The Benefits of Environmental Improvement.* Johns Hopkins University Press, Baltimore, Maryland.

Freeman, A.M. III (1982) *Air and Water Pollution Control: a Benefit–Cost Assessment.* John Wiley & Sons, New York.

Freeman, A.M. III (1990) Water pollution policy. In: Portney, P. (ed.) *Policies for Environmental Protection.* Resources for the Future, Washington, DC.

Freeman, A.M. III (1993) *The Measurement of Environmental and Resource Values: Theory and Method.* Resources for the Future, Washington, DC.

Freeman, A.M. III (1994) Clean Water Act Reauthorization: how far have we come? *Water Resources Bulletin* 30(5):793–798.

Gabel, H. (1994) The environmental impacts of trade in the transport sector. In: *The Environmental Effects of Trade.* Organisation for Economic Cooperation and Development, Paris.

Gardner, B.L. (1987) *The Economics of Agricultural Policies.* Macmillan Publishing, New York.

Gardner, K. and Barrows, R. (1985) The impact of soil conservation investments on land prices. *American Journal of Agricultural Economics* 67: 943–947.

Gardner, R.L. and Young, R.A. (1988) Assessing strategies for control of irrigation-induced salinity in the upper Colorado River basin. *American Journal of Agricultural Economics* 70:37–49.

Ghelfi, L., Gunderson, C., Johnson, J., Kassel, K., Kuhn, B., Mishrok, A., Morehart, M., Offutt, S., Tiehen, L. and Whitener, L. (2000) A safety net for farm households? *Agricultural Outlook* (January–February):19–24.

GLTN (Great Lakes Trading Network) (2000) Summary of program and project drivers and other presentation materials. *Markets for the New Millennium: How Can Water Quality Trading Work for You? Conference and Workshop*, Chicago, Illinois, 18–19 May.

Goldin, I., Knudsen, O. and Mensbrugghe, D. van der (1993) *Trade Liberalisation: Global Economic Implications*, Organisation for Economic Cooperation and Development, Paris.

Goolsby, D.A. and Battaglin, W.A. (1993) Occurrence, Distribution, and Transport of Agricultural Chemicals in Surface Waters of the Midwestern United States. In: Goolsby, D.A., Boyer, L.L. and Mallard, G.E. (eds), *Selected Papers on Agricultural Chemicals in Water Resources of the Midcontinental United States.* Open-File Report 93–418. US Department of Interior, US Geological Survey, Reston, Virginia, pp. 1–25.

Goolsby, D.A., Battaglin, W.A., Lawrence, G.B., Artz, R.S., Aulenbach, B.T., Hooper, R.P., Keeney, D.R. and Stensland, G.J. (1999) *Flux and Source of Nutrients in the Mississippi-Atchafalaya River Basin: Topic 3 Report.* Report submitted to White House Office of Science and Technology Policy, Committee on Environment and Natural Resources, Hypoxia Work Group, Silver Spring, Maryland, May.

Gould, B.W., Saupe, W.E. and Klemme, R.M. (1989) Conservation tillage: the role of farm and operator characteristics and the perception of soil erosion. *Land Economics* 85(2):167–182.

Govindasamy, R., Herriges, J.A. and Shogren, J.F. (1994) Nonpoint tournaments. In:

Dosi, C. and Tomasi, T. (eds) *Nonpoint Source Pollution Regulation: Issues and Analysis*. Kluwer Academic Press, Dordrecht, The Netherlands.

Graham, E. (1997) TMDLs and effluent trading: the key to healthy watersheds? Remarks before the conference *TMDL and Effluent Trading: The Key to Healthy Watersheds?*, Chesapeake Water Environment Association and the American Water Resources Association, College Park, Maryland, November.

Gren, I.-M., Elofsson, K. and Jannke, P. (1997) Cost effective nutrient reductions to the Baltic Sea. *Environmental and Resource Economics* 10:341–362.

Griffin, R.C. and Bromley, D.W. (1982) Agricultural runoff as a nonpoint externality: a theoretical development. *American Journal of Agricultural Economics* 64:37–49.

Grossman, G.M. (1995) Pollution and growth: what do we know? In: *The Economics of Sustainable Development*. Cambridge University Press, Cambridge, UK.

Grossman, G.M. and Helpman, E. (1991) *Innovation and Growth in the Global Economy*. MIT Press, Cambridge, Massachusetts.

Grossman, G.M. and Krueger, A. (1992) *Environmental Impacts of a North American Free Trade Agreement*. Discussion Paper, John M. Olin Program for the Study of Economic Organization and Public Policy, Princeton University, Princeton, New Jersey.

Hahn, R. (1989) Economic prescriptions for environmental problems: how the patient followed the doctor's orders. *Journal of Economic Perspectives* 3:95–114.

Hanley, N. (1999) Cost benefit analysis of environmental policy and management. In: Bergh, J. van den (ed.) *Handbook of Environmental and Resource Economics*. Edward Elgar, Cheltenham, UK.

Hanley, N. and Spash, C.L. (1993) *Cost Benefit Analysis and the Environment*. Edward Elgar, Aldershot, UK.

Hanley, N., Shogren, J.F. and White, B. (1997) *Environmental Economics in Theory and Practice*. Oxford University Press, New York.

Hanley, N., Kirkpatrick, H., Simpson, I. and Oglethorpe, D. (1998a) Principals for the provision of public goods from agriculture: modeling moorland conservation in Scotland. *Land Economics* 74(1):102–113.

Hanley, N., Wright, M. and Adamovicz, V. (1998b) Using choice experiments to value the environment. *Environmental and Resource Economics* 11:413–428.

Hanley, N., Whitby, M. and Simpson, I. (1999) Assessing the success of agri-environmental policy in the UK. *Land Use Policy* 16: 67–80.

Hansen, L.G. (1998) A damage based tax mechanism for regulation of non-point emissions. *Environmental and Resource Economics* 12(1):99–112.

Harford, J.D. (1984) Averting behavior and the benefits of reduced soiling. *Journal of Environmental Economics and Management* 11:296–302.

Harrington, W., Krupnick, A.J. and Peskin, H.M. (1985) Policies for nonpoint water pollution. *Journal of Soil and Water Conservation* 40:27–32.

Hayami, Y. and Ruttan, V.W. (1985) *Agricultural Development: an International Perspective*. Johns Hopkins University Press, Baltimore, Maryland.

Heberling, M.T. (2000) Three essays on valuing public goods using the stated choice method. PhD Dissertation, Pennsylvania State University, University Park, Pennsylvania.

Heimlich, R.E. (1994) Targeting green support payments: the geographic interface between agriculture and the environment. In: Lynch, S. (ed.) *Designing Greeen Support Programs*. Henry Wallace Institute for Alternative Agriculture, Greenbelt, Maryland.

Heimlich, R.E., Fernandez-Cornejo, J., McBride, W., Klotz-Ingram, C., Jans, S. and Brooks, N. (2000) Genetically engineered crops: has adoption reduced pesticide use? *Agricultural Outlook* (August):13–17.

Helfand, G.E. and House, B.W. (1995) Regulating nonpoint source pollution under heterogeneous conditions. *American Journal of Agricultural Economics*, 77:1024–1032.

Hellerstein, D. (1992) Estimating consumer surplus in the censored linear model. *Land Economics* 68:83–92.

Helming, J. (1997) Impacts of manure policies for The Netherlands. In: Brouwer, F. and Kleinhanss, W. (eds) *The Implementation of Nitrate Policies in Europe*. Wissenschaftsverlag Vuak Kiel, Kiel, Germany.

Herriges, J.R., Govindasamy, R. and Shogren, J. (1994) Budget-balancing incentive mechanisms. *Journal of Environmental Economics and Management* 27:275–285.

Hickman, J.S., Rowell, C.P. and Williams, J.R. (1989) Net returns from conservation compliance for a producer and landlord in northeastern Kansas. *Journal of Soil and Water Conservation* 44(5):532–534.

Hilton, F.G.H. and Levinson, A. (1998) Factoring the environmental Kuznets curve: evidence from automotive lead emissions. *Journal of Environmental Economics and Management* 35:126–141.

Hoag, D.L. and Holloway, H.A. (1991) Farm production decisions under cross and conservation compliance. *American Journal of Agricultural Economics* 73(1):184–193.

Hoag, D.L. and Hughes-Popp, J.S. (1997) Theory and practice of pollution credit trading in water quality management. *Review of Agricultural Economics* 19(2):252–262.

Hoban, T.J. and Wimberley, R.C. (1992) Farm operators' attitudes about water quality and the RCWP. In: *Seminar Publication: the National Rural Clean Water Program*. US Environmental Protection Agency, Washington, DC.

Hoehn, J.T. and Loomis, J. (1993) Substitution effects in the valuation of multiple environmental programs. *Journal of Environmental Economics and Management* 25:56–75.

Holden, P.W. (1986) *Pesticides and Groundwater Quality*. National Academy Press, Washington, DC.

Holmes, T. (1988) The offsite impact of soil erosion on the water treatment industry. *Land Economics* 64(4):356–366.

Hopkins, J., Schnitkey, G. and Tweeten, L. (1996) Impacts of nitrogen control policies on crop and livestock farms at two Ohio farm sites. *Review of Agricultural Economics* 18:311–324.

Horan, R.D. (2001a) Cost-effective and stochastic dominance approaches to stochastic pollution control. *Environmental and Resource Economics* (in press).

Horan, R.D. (2001b) Differences in social and public risk perceptions and conflicting impacts on point/nonpoint trading ratios. *American Journal of Agricultural Economics* (in press).

Horan, R.D., Shortle, J.S. and Abler, D.G. (1998) Ambient taxes when polluters have multiple choices. *Journal of Environmental Economics and Management* 36:186–199.

Horan, R.D., Shortle, J.S. and Abler, D.G. (1999a) Green payments for nonpoint pollution control. *American Journal of Agricultural Economics* 81:1210–1215.

Horan, R.D., Shortle, J.S. and Abler, D.G. (1999b) *Ambient Taxes Under m-Dimensional Choice Sets, Heterogeneous Expectations, and Risk-Aversion*. Working Paper,

Department of Agricultural Economics, Michigan State University, East Lansing, Michigan.

Horan, R.D., Shortle, J.S. and Abler, D.G. (2000a) *The Design and Comparative Economic Performance of Alternative Point/Nonpoint Trading Markets. II. An Empirical Investigation of Nutrient Trading in the Susquehanna River Basin.* Working Paper, Department of Agricultural Economics, Michigan State University, East Lansing, Michigan.

Horan, R.D., Shortle, J.S., Abler, D.G. and Ribaudo, M. (2000b) *The Design and Comparative Economic Performance of Alternative Point/Nonpoint Trading Markets. I. Theoretical Issues.* Working Paper, Department of Agricultural Economics, Michigan State University, East Lansing, Michigan.

Horan, R.D., Abler, D.G., Shortle, J.S., Carmichael, J. and Wang, L. (2001) Probabilistic, cost-effective point/nonpoint management in the Susquehanna River Basin. Paper presented at the Integrated Decision-Making for Watershed Management Symposium, Chevy Chase, Maryland, January 2001.

House, R.M., Peters, M. and McDowell, H. (1999) *USMP Regional Agricultural Model.* US Department of Agriculture, Economic Research Service, Washington, DC (unpublished), 88 pp.

Houston, J.E. and Henglun Sun (1999) Cost-share incentives and best management practices in a pilot water quality program. *Journal of Agricultural and Resource Economics* 24(1):239–252.

Howitt, R.E. (1995) Positive mathematical programming. *American Journal of Agricultural Economics* 77:329–342.

Hrubovcak, J., LeBlanc, M. and Miranowski, J. (1989) Limitations in evaluating environmental and agricultural policy coordination benefits. *American Economic Review* 80:208–212.

Hrubovcak, J., LeBlanc, M. and Eakin, B.K. (1995) *Accounting for the Environment in Agriculture.* TB-1847, October, US Department of Agriculture, Economic Research Service, Washington, DC.

Hrubovcak, J., Utpal Vasavada and Aldy, J.E. (1999) *An Economic Research Service Report: Green Technologies for a More Sustainable Agriculture.* Agricultural Information Bulletin Number 752, US Department of Agriculture, Economic Research Service, Washington, DC.

Huang, W. and Le Blanc, M. (1994) Market-based incentives for addressing non-point water quality problems: a residual nitrogen tax approach. *Review of Agricultural Economics* 16:427–440.

Huang, W.-Y., Shank, D. and Irwin Hewitt, T. (1996) On-farm costs of reducing residual nitrogen on cropland vulnerable to nitrate leaching. *Review of Agricultural Economics* 12:325–339.

Hubbell, B.J. and Carlson, G.A. (1998) Effects of insecticide attributes on within-season insecticide product and rate choices: the case of US apple growers. *American Journal of Agricultural Economics* 80(2):382–396.

Huffman, W.E. and Evenson, R.E. (1989) Supply and demand functions for multi-product US cash grain farms: biases caused by research and other policies. *American Journal of Agricultural Economics* 71:761–773.

Jacobs, J.J. and Casler, G.L. (1979) Internalizing externalities of phosphorus discharges from crop production to surface water: effluent taxes versus uniform reductions. *American Journal of Agricultural Economics* 60: 309–312.

Johansson, P. (1987) *The Economic Theory and Measurement of Environmental Benefits.* Cambridge University Press, Cambridge, UK.

Johnson, S.L., Adams, R.M. and Perry, G.M. (1991) The on-farm costs of reducing groundwater pollution. *American Journal of Agricultural Economics* 73:1063–1073.

Jung, C., Krutilla, K. and Boyd, R. (1996) Incentives for advanced pollution abatement technology at the industry level: an evaluation of policy alternatives. *Journal of Environmental Economics and Management* 30:95–111.

Juranek, D. (1995) *Cryptosporidiosis:* source of infection and guidelines for prevention. *Clinical Infectious Diseases* 21(Supplement 1):57–61.

Just, R.E. and Antle, J.M. (1991) Effects of commodity program structure on resource use and the environment. In: Just, R.E. and Bockstael, N. (eds) *Commodity and Resource Policies in Agricultural Systems.* Springer-Verlag, Berlin.

Just, R.E. and Rausser, G.C. (1992) Environmental and agricultural policy linkages and reforms in the United States under GATT. *American Journal of Agricultural Economics* 74:766–774.

Just, R.E., Hueth, D.L. and Schmitz, A. (1982) *Applied Welfare Economics and Public Policy.* Prentice Hall, Englewood Cliffs, New Jersey.

Kahn, F.R. and Kemp, W.M. (1985) Economic losses associated with the degradation of an ecosystem: the case of submerged aquatic vegetation in Chesapeake Bay. *Journal of Environmental Economics and Management* 12:246–263.

Kahneman, D. (1986) Comments. In: Cummings, R.G., Brookshire, D.S. and Schulze, W.D. (eds) *Valuing Environmental Goods.* Rowman and Allanheld, Totowa, New Jersey.

Kahneman, D. and Knetsch, J. (1992) Valuing public goods: the purchase of moral satisfaction. *Journal of Environmental Economics and Management* 22:57–70.

Kaoru, Y., Smith, V.K. and Liu, J.L. (1995) Using random utility models to estimate the recreational value of estuarine resources. *American Journal of Agricultural Economics*, 77:141–151.

Karr, J.R. and Chu, E.W. (1999) *Restoring Life in Running Waters: Better Biological Monitoring.* Island Press, Washington, DC.

Kellogg, R.L., Wallace, S., Alt, K. and Goss, D.W. (1997) Potential Priority Watersheds for Protection of Water Quality from Nonpoint Sources Related to Agriculture. 52nd Annual Soil and Water Conservation Society Conference, Toronto, Ontario.

Kemp, M.A. and Maxwell, C. (1992) Exploring a budget context for contingent valuation estimates. Paper presented at the Cambridge Economics, Inc. Symposium, *Contingent Valuation: A Critical Assessment,* Washington, DC, April.

Kemp, R., Olsthoorn, X., Oosterhuis, F. and Verbruggen, H. (1992) Supply and demand factors of cleaner technologies: some empirical evidence, *Environmental and Resource Economics* 2:615–634.

Kim, C.S., Hostetler, J. and Amacher, G. (1993) The regulation of groundwater quality with delayed responses. *Water Resources Research* 29:1369–1377.

Knisel, W.G. (1980) *CREAMS: a Field Scale Model for Chemicals, Runoff, and Erosion from Agricultural Management Systems.* Conservation Research Report No. 25, US Department of Agriculture, Agricultural Research Service, Washington, DC.

Knopman, D.S. and Smith, R.A. (1993) 20 years of the Clean Water Act. *Environment* 35(1):17–20, 35–51.

Knox, D., Jackson, G. and Nevers, E. (1995) Farm A Syst *Progress Report 1991–1994.* University of Wisconsin Cooperative Extension. *Conservation* 47(3):260–263.

Koop, G. and Tole, L. (1999) Is there an environmental Kuznets curve for deforestation? *Journal of Development Economics* 58:231–244.

Krueger, A.O., Schiff, M. and Valdés, A. (1992) *The Political Economy of Agricultural Pricing Policy*, Vols 1–5. Johns Hopkins University Press, Baltimore, Maryland.

Krutilla, F. (1999) Environmental policy and transaction costs. In: Vandenbergh, J. (ed.) *Handbook of Environmental and Resource Economics.* Edward Elgar, Cheltenham, UK.

Langemeier, R.N. (1992) Memo to J.W. Meck, USDA Working Group on Water Quality from Chief, Drinking Water Branch, US EPA Region VII, 29 April.

Larson, D., Helfand, G. and House, B. (1996) Second-best tax policies to reduce nonpoint source pollution. *American Journal of Agricultural Economics* 78(4):1108–1117.

Laughland, A.S. (1994) Multiple instruments in the regulation of nonpoint source pollution. PhD Dissertation, Department of Agricultural Economics and Rural Sociology, Pennsylvania State University, University Park, Pennsylvania.

Lee, D.J., Howitt, R.E. and Marino, M.A. (1993) A stochastic model of river water quality: application to salinity in the Colorado River. *Water Resources Research* 29(12):3917–3923.

Lee, J.G., Lacewell, R.D. and Richardson, J.W. (1991) Soil conservation or commodity programs: tradeoffs during the transition to dryland crop production. *Southern Journal of Agricultural Economics* 23(1): 203–211.

Leggett, C.G. and Bockstael, N.E. (2000) Evidence of the effects of water quality on residential prices. *Journal of Environmental Economics and Management* 39:121–144.

Leonard, R.A., Knisel, W.G. and Still, D.A. (1987) GLEAMS. Groundwater loading effects of agricultural management systems. *Transactions of the American Society of Agricultural Engineers* 30:1403–1418.

Letson, D. (1992) Point/nonpoint source trading: an interpretive survey. *Natural Resources Journal* 32:219–232.

Letson, D. and Gollehon, N. (1996) Confined animal production and the manure problem. *Choices* (3rd Quarter):18–24.

Letson, D., Crutchfield, S. and Malik, A. (1993) *Point/Nonpoint Source trading for Managing Agricultural Pollution Loadings.* US Department of Agriculture Economic Research Service, Agricultural Economics Report No. 674, US Government Printing Office, Washington, DC

Lettenmaier, D.P., Hooper, E.R., Wagoner, C. and Faris, K.B. (1991) Trends in stream quality in the continental United States, 1978–1987. *Water Resources Research* 27(3):327–339.

Leuck, D.J. (1993) *Policies to Reduce Nitrate Pollution in the European Community and Possible Effects on Livestock Production.* Staff Report No. AGES 9318, US Department of Agriculture, Economic Research Service, Washington, DC.

Liapis, P.S. (1994) Environmental and economic implications of alternative EC policies. *Journal of Agricultural and Applied Economics* 26:241–251.

Lichtenberg, E. (1992) Alternative approaches to pesticide regulation. *Northeastern Journal of Agricultural and Resource Economics* 21:83–92.

Lichtenberg, E. and Lessley, B.V. (1992) Water quality, cost-sharing, and technical assistance: perceptions of Maryland farmers. *Journal of Soil and Water Conservation* 47:260–263.

Lichtenberg, E. and Zilberman, D. (1986) The welfare economics of price supports in US agriculture. *American Economic Review* 76:1135–1141.

Lichtenberg, E. and Zilberman, D. (1988) Efficient regulation of environmental health risks. *Quarterly Journal of Economics* 49:167–168.

Lichtenberg, E., Zilberman, D. and Bogen, K.T. (1989) Regulating environmental health risks under uncertainty: groundwater contamination in Florida. *Journal of Environmental Economics and Management* 17:22–34.

Lipsey, R.G. and Lancaster, K. (1956) The general theory of second best. *Review of Economic Studies* 24:11–32.

Logan, T.J. (1993) Agricultural best management practices for water pollution control: current issues. *Agriculture, Ecosystems and Environment* 46:223–231.

Lohman, L.C., Milliken, J.G. and Dorn, W.S. (1988) *Estimating Economic Impacts of Salinity of the Colorado River.* US Department of Interior, Bureau of Reclamation, Denver, Colorado, February.

Loomis, J., Hanemenn, W.M., Kanninen, B. and Wegge, T. (1991) Willingness to pay to protect wetland and reduce wildlife contamination from agricultural drainage. In: Dinar, A. and Zilbelman, D. (eds) *The Economics and Management of Water and Drainage in Agriculture.* Kluwer Academic, Norwell, Massachusetts.

Lopes, L.L. (1992) Risk perception and the perceived public. In: Bromley, D. and Segerson, S. (eds) *The Social Response to Environmental Risk: Policy Formation in an Age of Uncertainty.* Kluwer Academic Publishers, Dordrecht, The Netherlands.

Low, P. (ed.) (1992) *International Trade and the Environment.* Discussion Paper No. 159, World Bank, Washington, DC.

Lucas, R.E.B. (1996) International environmental indicators: trade, income, and endowments. In: *Agriculture, Trade, and the Environment: Discovering and Measuring the Critical Linkages.* Westview Press, Boulder, Colorado.

Lynch, S. and Smith, K. (1994) *Lean, Mean, and Green ... Designing Farm Support Programs in a New Era.* Henry Wallace Institute for Alternative Agriculture, Greenbelt, Maryland.

MacKenzie, W.R., Hoxie, N.J., Proctor, M.E., Gradus, M.S., Blair, K.A., Peterson, D.E., Kazmeirczak, J.J., Addiss, D.G., Fox, K.R., Rose, J.B. and Davis, J.P. (1994) A massive outbreak in Milwaukee of *Cryptosporidium* infection transmitted through the public water supply. *New England Journal of Medicine* 331(3):161–167.

Maddala, G.S. (1983) *Limited-Dependent and Qualitative Variables in Econometrics.* Cambridge University Press, Cambridge, UK.

Magat, A.W. (1978) Pollution control and technological advance: a dynamic model of the firm. *Journal of Environmental Economics and Management* 5:1–25.

Magat, A.W. (1979) The effects of environmental regulation on innovation. *Law and Contemporary Problems* 43:4–25.

MAFF (1999) *Nitrate Sensitive Areas.* Accessed at www.maff.gov.uk/environ/envsch/nas.htm, July.

Malcomson, J.M. (1978) Prices vs. quantities: a critical note on the use of approximations. *Review of Economic Studies* 68:203–207.

Maler, K. (1974) *Environmental Economics: a Theoretical Inquiry.* Johns Hopkins University Press, Baltimore, Maryland.

Maler, K.-G. (1977) A note on the use of property values in estimating marginal willingness to pay for environmental quality. *Journal of Environmental Economics and Management* 4:355–369.

Malik, A.S., Letson, D. and Crutchfield, S.R. (1993) Point/nonpoint source trading of pollution abatement: choosing the right trading ratio. *American Journal of Agricultural Economics* 75:959–967.

Malik, A.S., Larson, B.A. and Ribaudo, M.O. (1994) Economic incentives for agricultural nonpoint pollution control. *Water Resources Bulletin* 30:471–480.

Mapp, H.P., Bernardo, D.J., Sabbagh, G.J., Geleta, S. and Watkins, K.B. (1994) Economic and environmental impacts of limiting nitrogen use to protect water quality: a stochastic regional analysis. *American Journal of Agricultural Economics* 76(4):889–903.

McCann, L. and Easter, K.W. (1999) Transaction costs of policies to reduce agricultural phosphorus pollution in the Minnesota River. *Land Economics* 75(3):402–414.

McConnell, K.E. (1983) An economic model of soil conservation. *American Journal of Agricultural Economics* 65: 83–89.

McConnell, K.E. and Strand, I. (1989) Benefits from commercial fisheries when demand and supply depend on water quality. *Journal of Environmental Economics and Management* 17:284–292.

McGuikin, J.T. and Young, R.A. (1981) On the economics of desalination of brackish water supplies. *Journal of Environmental Economics and Management* 8:79–91.

McSweeny, W.T. and Shortle, J.S. (1990) Probabilistic cost effectiveness in agricultural nonpoint pollution control. *Southern Journal of Agricultural Economics* 22:95–104.

Mendelsohn, R. (1984) Estimating the structural equations of implicit markets and household production functions. *Review of Economics and Statistics* 66:673–677.

Mendelsohn, R. and Brown, G.M. Jr (1983) Revealed preference approaches to valuing outdoor recreation. *Natural Resources Journal* 23:607–618.

Menell, P. (1990) The limitations of legal institutions for addressing environmental risk. *Journal of Economic Perspectives* 5:93–114.

Miceli, T. and Segerson, K. (1991) Joint liability in torts and infra-marginal efficiency. *International Review of Law and Economics* 11:235–249.

Milliman, R.S. and Prince, R. (1989) Firm incentives to promote technological change in pollution control. *Journal of Environmental Economics and Management* 17:247–265.

Milon, J.W. (1987) Optimizing nonpoint source controls in water quality. *Water Resources Bulletin* 23:387–396.

Miranowski, J.A. and Hammes, B.D. (1984) Implicit prices for soil characteristics in Iowa. *American Journal of Agricultural Economics* 66: 379–383.

Miranowski, J.A., Hrubovcak, J. and Sutton, J. (1991) The effects of commodity programs on resource use. In: Just, R.E. and Bockstael, N.E. (eds) *Commodity and Resource Policies in Agricultural Systems*. Springer Verlag, Berlin.

Mirvish, S.S. (1991) The significance for human health of nitrate, nitrite and *N*-nitroso compounds. In: Bogárdi, I. and Kuzelka, R.D. (eds) *Nitrate Contamination: Exposure, Consequence, and Control*. Springer-Verlag, Berlin, pp. 253–266.

Mitchell, R. and Carson, R. (1989) *Using Surveys to Value Public Goods: the Contingent Valuation Method*. Resources for the Future, Washington, DC.

Montgomery, W.D. (1972) Markets in licenses and efficient pollution control. *Journal of Economics Theory* 5:395–418.

Moore, M.R., Mulville, A. and Weinberg, M. (1996) Water allocation in the American west: endangered fish versus irrigated agriculture. *Natural Resource Journal* 36:319–357.

Morandi, L. (1989) *State Groundwater Protection Policies: a Legislator's Guide*. National Conference of State Legislatures, Washington, DC.

Moreau, D.H. (1995) Water pollution control in the United States: policies, planning and criteria. *Water Resources Update* 94:2–23.

Moreau, R. and Strasma, J. (1995) Measuring the benefits and costs of voluntary pollution prevention programs in the agricultural sector under three alternative concepts – averting expenditures, willingness-to-pay, and avoidance costs: a benefit–cost analysis of the farm assessment system (Farm*A*Syst) as implemented in nine parishes in Louisiana. Unpublished paper, University of Wisconsin.

Morey, E., Shaw, W.D. and Rowe, R. (1991) A discrete-choice model of recreational participation, site choice, and activity valuation when complete trip data are not available. *Journal of Environmental Economics and Management* 20:181–201.

Moxey, A. and White, B. (1994) Efficient compliance with agricultural nitrate pollution standards. *Journal of Agricultural Economics* 45(1):27–37.

Mueller, D.K. and Helsel, D.R. (1996) *Nutrients in the Nation's Water – Too Much of a Good Thing?* US Geological Survey Circular 1136. US Department of Interior, US Geological Survey, Denver, Colorado.

Mueller, D.K., Hamilton, P.A., Helsel, D.R., Hitt, K.J. and Ruddy, B.C. (1995) *Nutrients in Ground Water and Surface Water of the United States – an Analysis of Data Through 1992.* Water-Resources Investigations Report 95–4031. US Department of Interior, US Geological Survey, Reston, Virginia.

Mundlak, Y. (1970) Further implications of distortion in the factor market. *Econometrica.* 38:517–532.

Musser, W., Shortle, J.S., Kreahling, K., Roach, B., Huang, W., Beegle, D. and Fox, R.H. (1995) An economic analysis of corn nitrogen test for Pennsylvania corn. *Review of Agricultural Economics* 17:25–35.

Napier, T.L. and Brown, D.E. (1993) Factors affecting attitudes toward groundwater pollution among Ohio farmers. *Journal of Soil and Water Conservation* 48(5):432–438.

Napier, T.L. and Camboni, S.M. (1993) Use of conventional and conservation practices among farmers in the Scioto River Basin of Ohio. *Journal of Soil and Water Conservation* 48(3):231–237.

National Research Council (1989) *Alternative Agriculture.* National Academy Press, Washington, DC.

National Research Council (1993) *Soil and Water Quality: an Agenda for Agriculture.* National Academy Press, Washington, DC.

National Research Council (2000) *Clean Coastal Waters: Understanding and Reducing the Effects of Nutrient Pollution.* Committee on the Causes and Management of Eutrophication, Ocean Studies Board and Water Science and Technology Board. National Academy Press, Washington, DC.

National Science and Technology Council (2000) *An Integrated Assessment: Hypoxia in the Northern Gulf of Mexico.* Committee on Environment and Natural Resources, Washington, DC, 58 pp.

Nelson, M.C. and Seitz, W.D. (1979) An economic analysis of soil erosion control in a watershed representing corn belt conditions. *North Central Journal of Agricultural Economics* 1(2):173–186.

Nguyen, T., Perroni, C. and Wigle, R. (1993) An evaluation of the draft Final Act of the Uruguay Round. *Economic Journal* 103:1540–1549.

Nielsen, E.G. and Lee, L.K. (1987) *The Magnitude and Cost of Groundwater Contamination from Agricultural Chemicals – a National Perspective.* Agricultural Economics Report No. 576. US Department of Agriculture, Economic Research Service. Government Printing Office, Washington, DC.

Norton, N.A., Phipps, T.T. and Fletcher, J.J. (1994) Role of voluntary programs in agricultural nonpoint pollution policy. *Contemporary Economic Policy* 12:113–121.

Novotny, V. and Olem, H. (1994) *Water Quality: Prevention, Identification, and Management of Diffuse Pollution.* Van Nostrand Reinhold, New York.

Nowak, P.J. (1987) The adoption of agricultural conservation technologies: economic and diffusion explanations. *Rural Sociology* 49(5):477–483, 52(2): 208–220.

Nowak, P.J., O'Keefe, G., Bennett, C., Anderson, S. and Trumbo, C. (1997) *Communication and Adoption Evaluation of USDA Water Quality Demonstration Projects.* University of Wisconsin in cooperation with US Department of Agriculture, Madison, Wisconsin.

NRAES (1996) *Animal Agriculture and the Environment: Nutrients, Pathogens, and Community Relations.* Proceedings from the Animal Agriculture and the Environment North American Conference, December. Northeast Regional Agricultural Engineering Service. Cooperative Extension, Ithaca, New York.

NRDC (1998) *America's Animal Factories: How States Fail to Prevent Pollution from Livestock Waste.* Natural Resources Defense Council. http://www.nrdc.org/nrdc/nrdcpro/factor/

Oates, W. (1995) Green taxes: can we protect the environment and improve the tax system at the same time? *Southern Economic Journal* 6(14):915–922.

OECD (1986) *Water Pollution by Fertilizers and Pesticides.* Organisation for Economic Cooperation and Development, Paris.

OECD (1989) *Agricultural and Environmental Policies: Opportunity for Integration.* Organisation for Economic Cooperation and Development, Paris.

OECD (1991) *State of the Environment.* Organisation for Economic Cooperation and Development, Paris.

OECD (1993a) *Agricultural and Environmental Policies Integration: Recent Progress and New Directions.* Organisation for Economic Cooperation and Development, Paris.

OECD (1993b) *Environmental Performance Reviews: Germany.* Organisation for Economic Cooperation and Development, Paris.

OECD (1994a) *The Environmental Effects of Trade.* Organisation for Economic Cooperation and Development, Paris.

OECD (1994b) *Methodologies for Environmental and Trade Reviews.* Organisation for Economic Cooperation and Development, Paris.

OECD (1994c) *Environmental Taxes in OECD Countries.* Organisation for Economic Cooperation and Development, Paris.

OECD (1994d) *Towards Sustainable Agricultural Production: Cleaner Technologies.* Organisation for Economic Cooperation and Development, Paris.

OECD (1995) *Environmental Performance Reviews: Netherlands.* Organisation for Economic Cooperation and Development, Paris.

OECD (1997) *Environmental Taxes and Green Tax Reform.* Organisation for Economic Cooperation and Development, Paris.

OECD (1998) *Agriculture and the Environment Issues and Policies.* Organisation for Economic Cooperation and Development, Paris.

OECD (1999) *Environmental Data Compendium.* Organisation for Economic Cooperation and Development, Paris.

OECD (2000) *Agricultural Policies in OECD Countries 2000: Monitoring and Evaluation.* Organisation for Economic Cooperation and Development, Paris.

Olson, E. (1995) *You Are What You Drink ...* Briefing paper, Natural Resources Defense Council, New York, June.

Opshoor, J.B., Savornin Lohman, A.F. de and Vos, H.B. (1994) *Managing the Environment: the Role of Economic Instruments.* Organisation for Economic Cooperation and Development, Paris.

Osborn, C.T. and Setia, P.P. (1988) Estimating the economic impacts of conservation compliance: a corn belt application. Selected paper presented at the American Agricultural Economics Association annual meeting, August, Knoxville, Tennessee.

Osteen, C.D. and Szmedra, P.I. (1989) *Agricultural Pesticide Use Trends and Policy Issues.* Agricultural Economics Report No. 622, US Department of Agriculture, Economic Research Service. Government Printing Office, Washington, DC.

Osterkamp, W.R., Heilman, P. and Lane, L.J. (1998) Economic considerations of a continental sediment-monitoring program. *International Journal of Sediment Research* 13:12–24.

Padgitt, S. (1989) *Farmers' Views on Groundwater Quality: Concerns, Practices, and Policy Preferences.* Staff report, February. US Office of Technology Assessment, Washington, DC.

Pait, A., DeSouza, A. and Farrow, D. (1992) *Agricultural Pesticide Use in Coastal Areas: a National Summary.* US Department of Commerce, National Oceanic and Atmospheric Administration, Washington, DC, September.

Palmquist, R.B. and Danielson, L.E. (1989) A hedonic study of the effects of erosion control and drainage on farmland values. *American Journal of Agricultural Economics* 71(1):55–62.

Park, W.M. and Shabman, L.A. (1982) Distributional constraints on acceptance of nonpoint pollution controls. *American Journal of Agricultural Economics* 64:455–462.

Pease, J. and Bosch, D. (1994) Relationships among farm operators' water quality opinions, fertilization practices, and cropland potential to pollute in two regions of Virginia. *Journal of Soil and Water Conservation* 49(5):477–483.

Petterrson, O. (1994) Reduced pesticide use in Scandinavian agriculture. *Critical Reviews in Plant Sciences* 13:43–55.

Phipps, T.T. (1991) Commercial agriculture and the environment: an evolutionary perspective. *Northeastern Journal of Agricultural and Natural Resource Economics* 20:143–150.

Pimentel, D. (ed.) (1993) *World Soil Erosion and Conservation.* Cambridge University Press, Cambridge, UK.

Pimentel, D., Andow, D., Dyson-Hudson, R., Gallahan, D., Jacobson, S., Irish, M., Kroop, S., Moss, A., Schreiner, I., Shepard, M., Thompson, T. and Vinzant, B. (1991) Environmental and social costs of pesticides: a preliminary assessment. In: Pimentel, D. (ed.) *Handbook of Pest Management in Agriculture*, Vol. I. CRC Press, Boca Raton, Florida, pp. 721–740.

Plantinga, A.J. (1996) The effect of agricultural policies on land use and environmental quality. *American Journal of Agricultural Economics* 78:1082–1091.

Poe, G.L. (1997) Extra-Market Values and Conflicting Agriculture Policies. *Choices* 1997 3rd quarter:4–8.

Poe, G.L. and Bishop, R.C. (1992) Measuring the benefits of groundwater protection from agricultural contamination: results from a two-stage contingent valuation study. Staff Paper No. 341, Department of Agricultural Economics, University of Wisconsin, Madison, Wisconsin.

Polinsky, M. and Rubinfeld, D. (1977) Property values and the benefits of environmental improvement: theory and measurement. In: *Public Economics and the Quality of Life.* Johns Hopkins University Press, Baltimore, Maryland.

Porter, M.E. and van der Linde, C. (1995) Toward a new conception of the environment–competitiveness relationship. *Journal of Economic Literature* 9:97–118.

Postel, S. (1999) *Pillar of Sand: Can the Irrigation Miracle Last?* W.W Norton and Co., London, New York.

Powell, J.R. (1991) The value of groundwater protection: measurement of willingness-to-pay, information and its utilization by local government decision makers. PhD Thesis, Department of Agricultural Economics, Cornell University.

Prato, T. and Wu, S. (1991) Erosion, sediment and economic effects of conservation compliance in an agricultural watershed. *Journal of Soil and Water Conservation* 46(3):211–214.

Preston, S.D. and Brakebill, J.W. (1999) *Application of Spatially Referenced Regression Modeling for the Evaluation of Total Nitrogen Loading in the Chesapeake Bay Watershed.* Water-Resources Investigations Report 99–4054, US Geological Survey, Reston, Virginia.

Puckett, L.J. (1995) Identifying major sources of nutrient water pollution. *Environmental Science and Technology* 29:408A–414A.

Qiu Zeyuan, Prato, A. and Kaylen, M. (1998) Watershed-scale economic and environmental tradeoffs incorporating risks. *Agricultural Resource Economics Review* 27:231–240.

Randall, A., Hoehn, J.P. and Brookshire, D.S. (1983) Contingent valuation surveys for evaluating environmental assets. *Natural Resources Journal* 23:635–648.

Randhir, T.O. and Lee, J.G. (1997) Economic and water quality impacts of reducing nitrogen and pesticide use in agriculture. *Agricultural and Resource Economics* 1997:39–51.

Ready, R.C. and Henken, K. (1999) Optimal self-protection from nitrate-contaminated groundwater. *American Journal of Agricultural Economics* 81(2): 321–334.

Rendleman, C.M., Reinert, K.A. and Tobey, J.A. (1995) Market-based systems for reducing chemical use in agriculture in the United States. *Environmental and Resource Economics* 5:51–70.

Ribaudo, M.O. (1986) *Reducing Soil Erosion: Offsite Benefits.* AER 561. US Department of Agriculture, Economic Research Service, Washington, DC, September.

Ribaudo, M.O. (1989) *Water Quality Benefits from the Conservation Reserve Program.* AER 606. US Department of Agriculture, Economic Research Service, Washington, DC, February.

Ribaudo, M.O. and Bouzaher, A. (1994) *Atrazine: Environmental Characteristics and Economics of Management.* AER-699. US Department of Agriculture, Economic Research Service, Washington, DC, September.

Ribaudo, M.O. and Horan, R.D. (1999) The role of education in nonpoint source pollution control policy. *Review of Agricultural Economics* 21:331–343.

Ribaudo, M.O., Horan, R.D. and Smith, M.E. (1999) *Economics of Water Quality Protection From Nonpoint Sources: Theory and Practice.* ERS Agricultural Economics Report AER-782. US Department of Agriculture, Economic Research Service, Resource Economics Division, Washington, DC.

Richardson, J.W., Gerloff, D.C., Harris, B.L. and Dollar, L.L. (1989) Economic impacts of conservation compliance on a representative Dawson County, Texas farm. *Journal of Soil and Water Conservation* 44(5):527–531.

Rigby, D. and Young, T. (1996) European environmental regulations to reduce water pollution: an analysis of their impact on UK dairy farms. *European Review of Agricultural Economics* 23(1):59–78.

Roberts, M. and Spence, M. (1976) Effluent charges and licenses under uncertainty. *Journal of Public Economics* 5:193–208.

Romstad, E. (1997) *Team Approaches in Reducing Nonpoint Source Pollution*. IOS-Discussion Paper D-01/1997, Department of Economics and Social Sciences, Agricultural University of Norway, Ås, Norway.

Rosen, S. (1974) Hedonic prices and implicit markets: product differentiation in pure competition. *Journal of Political Economy* 82:34–55.

Russell, C.S. and Powell, P.T. (2000) Practical considerations and comparison of instruments of environmental policy. In: Berg, J. van den (ed.) *Handbook of Environmental and Resource Economics*. Edward Elgar Publishing Inc., Cheltenham, UK.

Russell, C.S., Harrington, W. and Vaughan, W.J. (1986) *Enforcing Pollution Control Laws*. Resources for the Future Inc., Washington, DC.

Samples, K.C., Gowen, M.M. and Dixon, J.A. (1986) The validity of the contingent valuation method for estimating non-use components of preservation values for unique natural resources. Paper presented at the annual meeting of the American Agricultural Economics Association, Reno, Nevada, July.

Samuelson, P.A. (1970) The fundamental approximation theorem of portfolio analysis in terms of means, variances, and higher moments. *Review of Economic Studies* 37:537–542.

Sandmo, A. (1975) Optimal taxation in the presence of externalities. *Swedish Journal of Economics* 77(1): 86–98.

Schmitz, A., Boggess, W.G. and Tefertiller, K. (1995) Regulations: evidence from the Florida dairy industry. *American Journal of Agricultural Economics* 77(5):1166–1171.

Schneider, S.A. (1990) The regulation of agricultural practices to protect groundwater quality: the Nebraska model for controlling nitrate contamination. *Virginia Environmental Law Journal* 10:1–44.

Schnitkey, G.D. and Miranda, M. (1993) The impacts of pollution controls on livestock–crop productions. *Journal of Agricultural and Resource Economics* 18:25–36.

Schott, J.J. (1994) *The Uruguay Round: an Assessment*. Institute for International Economics, Washington, DC.

Schou, J. (1997) Implementation of nitrate policies in Denmark. In: Brouwer, F. and Kleinhanss, W. (eds) *The Implementation of Nitrate Policies in Europe*. Wissenschaftsverlag Vuak Kiel, Kiel, Germany.

Segerson, K. (1988) Uncertainty and incentives for non-point source pollution. *Journal of Environmental Economics and Management*. 15:87–98.

Segerson, K. (1990) Liability for groundwater contamination from pesticides. *Journal of Environmental Economics and Management* 19:227–243.

Segerson, K. (1995) Liability and penalty structures in policy design. In: Bromley, D.W. (ed.) *The Handbook of Environmental Economics*. Basil Blackwell, Cambridge, Massachusetts.

Segerson, K. (1996) Issues in the choice of environmental policy instruments. In: Braden, J., Folmer, H. and Ulen, T. (eds) *Environmental Policy with Political and Economic Integration*. Edward Elgar Publishing Inc., Cheltenham, UK.

Segerson, K. (1999) Flexible incentives: a unifying framework for policy analysis. In: Casey, F., Schmitz, A., Swinton, S. and Zilberman, D. (eds) *Flexible Incentives for the Adoption of Environmental Technologies in Agriculture*. Kluwer Academic Press, Norwell, Massachusetts.

Seip, K. and Strand, J. (1990) Willingness to pay for environmental goods in Norway: a contingent valuation study with real payment. Paper prepared for the SAF Center for Applied Research, Department of Economics, University of Oslo, 26 pp.

Shaffer, M.J., Halvorson, A.D. and Pierce, F.J. (1991) Nitrate leaching and economic analysis package (NLEAP): model description and application. In: Follett, R.F., Keeney, D.R. and Cruse, R.M. (eds) *Managing Nitrogen for Groundwater Quality and Farm Profitability*. Soil Science Society of America, Madison, Wisconsin, pp. 285–322.

Shavell, S. (1987) Liability versus other approaches to the control of risk. In: *Economic Analysis of Accident Law*. Harvard University Press, Cambridge, Massachusetts.

Shogren, J.F. (1993) Reforming nonpoint pollution policy. In: Russell, C.S. and Shogren, J.F. (eds) *Theory, Modeling and Experience in the Management of Nonpoint-Source Pollution*. Kluwer Academic Publishers, Dordrecht, The Netherlands, pp. 329–345.

Shortle, J.S. (1987) Allocative implications of comparisons between the marginal costs of point and nonpoint source pollution abatement. *Northeastern Journal of Agricultural and Resource Economics* 16:17–23.

Shortle, J.S. (1990) The allocative efficiency implications of water pollution abatement cost comparisons. *Water Resources Research* 26(5):793–797.

Shortle, J.S. (1995) Environmental federalism: the case of US agriculture? In: Braden, J.B., Folmer, H. and Ulen, T. (eds) *Environmental Policy with Economic and Political Integration: the European Union and the United States*. Edward Elgar, Cheltenham, UK.

Shortle, J.S. (1996) Environmental federalism and the control of water pollution from US agriculture: is the current allocation of responsibilities between national and local authorities about right? In: Braden, J.B., Folmer, H. and Ulen, T.S. (eds) *Environmental Policy with Political and Economic Integration*. Edward Elgar, Cheltenham, UK.

Shortle, J.S. and Abler, D.G. (1994) Incentives for agricultural nonpoint pollution control. In: Graham-Tomasi, T. and Dosi, C. (eds) *The Economics of Nonpoint Pollution Control: Theory and Issues*. Kluwer Academic Press, Dordrecht, The Netherlands.

Shortle, J.S. and Abler, D.G. (1997) Nonpoint pollution. In: Folmer, H. and Tietenberg, T. (eds) *International Yearbook of Environmental and Natural Resource Economics*. Edward Elgar, Cheltenham, UK.

Shortle, J.S. and Abler, D.G. (1999) Agriculture and the environment. In: Vandenbergh, J. (ed.) *Handbook of Environmental and Resource Economics*. Edward Elgar, Cheltenham, UK.

Shortle, J.S. and Dunn, J.W. (1986) The relative efficiency of agricultural source water pollution control policies. *American Journal of Environmental Economics* 68(3):688–677.

Shortle, J.S. and Dunn, J.W. (1991) Economics of control of nonpoint pollution from agriculture. In: Hanley, N. (ed.) *Farming and the Countryside: an Economic Analysis of Costs and Benefits*. CAB International, Wallingford, UK.

Shortle, J.S. and Griffin, R.C. (2000) *Irrigated Agriculture and the Environment*. Edward Elgar, Cheltenham, UK.

Shortle, J.S. and Laughland, A. (1994) Impacts of taxes to reduce agrichemical use when farm policy is endogenous. *Journal of Agricultural Economics* 45(1):3–14.

Shortle, J.S. and Miranowski, J.A. (1987) Intertemporal soil resource use: is it socially excessive? *Journal of Environmental Economics and Management* 14:99–111.

Shortle, J., Musser, W., Huang, W., Roach, B., Kreahling, K., Beegle, D. and Fox, R. (1992) Economic and environmental potential of the pre-sidedressing soil nitrate test. EPA Contract No. CR-817370–01–0. US Environmental Protection Agency.

Shortle, J.S., Horan, R.D. and Abler, D.G. (1998) Research issues in nonpoint pollution control. *Environmental and Resource Economics* 11(3/4):571–585.

Shortle, J.S., Faichney, R., Hanley, N. and Munro, A. (1999) Least-cost pollution allocations for probabilistic water quality targets to protect salmon on the Forth Estuary. In: Sorrel, S. and Skea, J. (eds) *Pollution for Sale: Emissions Trading and Joint Implementation*. Edward Elgar, Northampton, Massachusetts.

Shumway, R. (1995) Recent duality contributions in production economics. *Journal of Agricultural and Resource Economics* 20:178–194.

Shumway, C.R. and Chesser, R.R. (1994) Pesticide tax, cropping patterns, and water quality in South Central taxes. *Journal of Agricultural and Applied Economics* 26(1):224–240.

Smith, K.R. (1995) Time to 'green' US farm policy. *Issues in Science and Technology* 1:71–78.

Smith, M.E. and Ribaudo, M.O. (1998) The new Safe Drinking Water Act: implications for agriculture. *Choices* (3rd quarter):26–30.

Smith, R.A., Alexander, R.B. and Lanfear, K.J. (1993) Stream water quality in the conterminous United States – status and trends of selected indicators during the 1980s. In: *National Water Summary 1990–91*. Water Supply Paper 2400. US Department of Interior, US Geological Survey, Reston, Virginia, pp. 111–140.

Smith, R.B.W. and Tomasi, T.D. (1995) Transaction costs and agricultural nonpoint-source water pollution control policies. *Journal of Agricultural and Resource Economics* 20:277–290.

Smith, R.R.W. (1995b) The Conservation Reserve Program as a least-cost land retirement mechanism. *American Journal of Agricultural Economics* 77(Feb):93–105.

Smith, S.J., Sharpley, A.N. and Ahuja, L.R. (1993) Agricultural chemical discharge in surface water runoff. *Journal of Environmental Quality* 22:474–480.

Smith, V.K. (1992) Environmental costing for agriculture: will it be standard fare in the Farm Bill of 2000? *American Journal of Agricultural Economics* 74:1076–1088.

Smith, V.K. (1997) Pricing what is priceless: a status report on nonmarket valuation. In: Folmer, H. and Tietenberg, T. (eds) *The International Yearbook of Environmental and Resource Economics 1997/1998: A Survey of Current Issues*. Edward Elgar, Cheltenham, UK.

Smith, V.K. and Desvousges, W. (1985) The generalized travel cost model and water quality benefits: a reconsideration. *Southern Economic Journal* 52:371–381.

Smith, V.K. and Kaoru, Y. (1987) The hedonic travel cost method: a view from the trenches. *Land Economics* 63:179–192.

Spalding, R.F. and Exner, M.E. (1993) Occurrence of nitrate in groundwater – a review. *Journal of Environmental Quality* 22:392–402.

Spash, C. and Falconer, K. (1997) Agri-environmental policies: cross-achievement and the role for cross-compliance. In: Brouwer, F. and Kleinhanss, W. (eds) *The Implementation of Nitrate Policies in Europe*. Wissenschaftsverlag Vuak Kiel, Kiel. Germany.

Stavins, R.N. (1996) Correlated uncertainty and policy instrument choice. *Journal of Environmental Economics and Management* 30:218–232.

Stewart, L., Hanley, N. and Simpson, I. (1997) *Economic Evaluation of Agri-environmental Schemes in the UK*. Report to HM Treasury and MAFF. Environmental Economics Research Group, University of Stirling, UK.

Stokes, C.S. and Brace, K.D. (1988) Agricultural chemical use and cancer mortality in selected rural counties in the USA. *Journal of Rural Studies* 3:239–247.

Strutt, A. and Anderson, K. (1999) Estimating environmental effects of trade agreements with global CGE models: a GTAP application to Indonesia. Paper presented at OECD Workshop on Methodologies for Environmental Assessments of Trade Liberalization Agreements. http://www.oecd.org/ech/26–27oct/docs/strutt.pdf, accessed March 2000.

Sumelius, J. (1997) Improvement of agri-environmental programmes for landscape and biodiversity. Paper to Cultural Landscape Symposium, Norges landbruk-shogskole, Ås, Norway.

Swinton, S.M. and Clark, D.S. (1994) Farm-level evaluation of alternative policy approaches to reduce nitrate leaching from Midwest agriculture. *Agricultural and Resource Economic Review,* 23:66–74.

Tanton, T.W. and Heaven, S. (1999) Worsening of the Aral Basin Crisis: can there be a solution? *Journal of Water Resources Planning and Management* 125:363–368.

Taylor, M.L., Adams, R.M. and Miller, S.F. (1992) Farmlevel response to agricultural effluent control strategies: the case of the Willamette Valley. *American Journal of Agricultural Economics* 17(1):173–185.

Teague, M.L., Bernardo, D.J. and Mapp, H.P. (1995) Farm level economic analysis incorporating stochastic environmental risk assessment. *American Journal of Agricultural Economics* 77(1):8–19.

Thompson, L.C., Atwood, J.D., Johnson, S.R. and Robertson, T. (1989) National implications of mandatory conservation compliance. *Journal of Soil and Water Conservation* 44(5):517–520.

Tietenberg, T.H. (1995a) Tradeable permits for pollution control when emission location matters: what have we learned? *Environmental and Resource Economics* 5:95–113.

Tietenberg, T.H. (1995b) Transferable discharge permits and global warming. In: Bromley, D.W. (ed.) *The Handbook of Environmental Economics.* Basil Blackwell, Cambridge, Massachusetts.

Tomasi, T., Segerson, K. and Braden, J. (1994) Issues in the design of incentive schemes for nonpoint source pollution control. In: Dosi, C. and Tomasi, T. (eds) *Nonpoint Source Pollution Regulation: Issues and Analysis.* Kluwer Academic Publishers, Dordrecht, The Netherlands, pp. 1–37.

Torras, M. and Boyce, J.K. (1998) Income, inequality, and pollution: a reassessment of the environmental Kuznets curve. *Ecological Economics* 25:147–160.

Trachtman, J.P. (1999) Assessment of the effects of trade liberalization on domestic environmental regulation: toward trade–environment policy integration. Paper presented at OECD Workshop on Methodologies for Environmental Assessments of Trade Liberalization Agreements. http://www.oecd.org/ech/26–27oct/docs/trachtman.pdf

UN Statistical Office (various years) *National Accounts Statistics: Main Aggregates and Detailed Tables.* United Nations, New York.

USDA (1993) *The USDA Water Quality Program Plan (1989).* Waterfax 000. US Department of Agriculture, Working Group on Water Quality, April.

USDA APHIS (1994) *Cryptosporidium and Giardia in Beef Calves.* National Animal Health Monitoring System report. US Department of Agriculture, Animal and Plant Health Inspection Service, January.

USDA ERS (1994) *Agricultural Resources and Environmental Indicators.* Agricultural Handbook No. 705. US Department of Agriculture, Economic Research Service, December.

USDA ERS (1997) *Agricultural Resources and Environmental Indicators.* Agricultural Handbook No. 712. US Department of Agriculture, Economic Research Service, July.

USDA NRCS (1996) *1994 Final Status Review Results* (unpublished). US Department of Agriculture, Natural Resources Conservation Service.

USDA and USEPA (1999) *Unified National Strategy for Animal Feeding Operations.* US Department of Agriculture and US Environmental Protection Agency, March.

USDI (2000) Central Valley Project Improvement Act – Public Law 102–575, Title 34. US Department of Interior, Bureau of Reclamation Mid-Pacific Region. http://www.mp.usbr.government/regional/cvpiamaim/index.html, accessed 14 June 2000.

USEPA (1984) *Report to Congress: Nonpoint Source Pollution in the US Office of Water Program Operations.* US Environmental Protection Agency, January.

USEPA (1988) *Nonpoint Sources: Agenda for the Future.* US Environmental Protection Agency, Office of Water, October.

USEPA (1992a) *Another Look: National Survey of Pesticides in Drinking Water Wells, Phase II Report.* EPA-579/09-91-020. US Environmental Protection Agency, January.

USEPA (1992b) *National Water Quality Inventory: 1990 Report to Congress.* EPA-503-9-92-006. US Environmental Protection Agency, April.

USEPA (1993) *Guide to Federal Water Quality Programs and Information.* EPA-230-B-93-001. US Environmental Protection Agency, Office of Policy, Planning, and Evaluation, February.

USEPA (1994a) *President Clinton's Clean Water Initiative: Analysis of Benefits and Costs.* EPA-800-R-94-002. US Environmental Protection Agency, Office of Water, March.

USEPA (1994b) *Drinking Water Regulations and Health Advisories.* US Environmental Protection Agency, Office of Water, May.

USEPA (1994c) *The Quality of Our Nation's Water: 1992.* EPA-841-S-94-002. US Environmental Protection Agency, Office of Water, March.

USEPA (1995) *National Water Quality Inventory: 1994 Report to Congress.* EPA-841-R-95-005. US Environmental Protection Agency, Office of Water, December.

USEPA (1996) *Waquoit Bay Watershed. Ecological Risk Assessment Planning and Problem Formulation* (draft). EPA-630-R-96-004a. US Environmental Protection Agency, Risk Assessment Forum, Washington, DC.

USEPA (1997a) *Drinking Water Infrastructure Needs Survey. First Report to Congress.* EPA--812-R-97–001. US Environmental Protection Agency, January.

USEPA (1997b) National Primary Drinking Water Regulations: Interim Enhanced Surface Water Treatment Rule Notice of Data Availability; Proposed Rule. *Federal Register* 3 November, pp. 59486–59557.

USEPA (1997c) *State Source Water Assessment and Protection Programs Guidance.* EPA-816-R-97-009. US Environmental Protection Agency, Office of Water, August.

USEPA (1997d) *TMDL (Total Maximum Daily Load) case studies.* US Environmental Protection Agency. http://www.epa.gov/OWOW/TMDL/docs.html

USEPA (1998a) *Drinking Water Contaminant List.* EPA-815-F-98-002. US Environmental Protection Agency, Office of Water, February.

USEPA (1998b) *National Water Quality Inventory: 1996 Report to Congress.* EPA-841-R-97-008. US Environmental Protection Agency, Office of Water, April.

USEPA (1998c) *Clean Water Action Plan: Restoring and Protecting America's Waters.* US Environmental Protection Agency, February.

USEPA (1998d) National Primary Drinking Water Regulation: Consumer Confidence Reports; Final Rule. *Federal Register* 63(160):44511–44536.

USEPA (1998e) *Wellhead Program and State Ground Water Protection Programs (CSG-WPP).* US Environmental Protection Agency, Office of Ground Water and Drinking Water. http://www.epa.gov/OGWDW/csgwell.htm

USEPA (2000a) *The Quality of Our Nation's Waters: a Summary of the National Water Quality.* US Environmental Protection Agency Office of Water. EPA-841-5-00-001.

USEPA (2000b) *Atlas of America's Polluted Waters.* EPA-840-B-00-002. US Environmental Protection Agency, Office of Water, May.

USEPA (2000c) *Guidelines for Preparing Economic Analyses.* EPA-240-R-00-003. US Environmental Protection Agency.

USGAO (1991a) EPA's Use of Benefit Assessment in Regulating Pesticides. GAO/RCED-9-52. US General Accounting Office. March.

USGAO (1991b) *Need for Greater EPA Leadership in Controlling Nonpoint Source Pollution.* GAO/T-RCED-91-60. US General Accounting Office, June.

USGS (1997) *Pesticides in Surface and Ground Water of the United States: Preliminary Results of the National Water Quality Assessment Program (NAWQA).* US Department of Interior, US Geological Survey, August.

USGS (1999) *The Quality of Our Nation's Waters: Nutrients and Pesticides.* US Geological Survey Circular 1225.

USGS (1999) USGS scientists tracking environmental damage from Floyd ... heavy flooding caused heavy pollution. News release 23 September 1999, Reston, Virginia.

US Office of Technology Assessment (1995) *Targeting Environmental Priorities in Agriculture: Reforming Program Strategies.* US Government Printing Office, OTA-ENV, Washington, DC.

Vajda, S. (1972) *Probabilistic Programming.* Academic Press, New York.

VanDyke, L.S., Pease, J.W., Bosch, D.J. and Baker, J. (1999) Nutrient management planning on four Virginia livestock farms: impacts on net income and nutrient losses. *Journal of Soil Water Conservation* 54(2):499–505.

Vasavada, U. and Warmerdam, S. (1998) Environmental policy and the WTO: unresolved questions. *Agricultural Outlook* (November):12–14.

Vatn, A., Bakken, L.R., Lundeby, H., Romstad, E., Rørstad, P. and Vold, A. (1997) Regulating nonpoint-source pollution from agriculture: an integrated modeling analysis. *European Review of Agricultural Economics* 24(2):207–229.

Vaughan, W. and Russell, C.S. (1982) Valuing a fishing day: application of a systematic varying parameter model. *Land Economics* 58:450–463.

Vincent, J.R. (1997) Testing for environmental Kuznets curves within a developing country. *Environment and Development Economics* 2:417–431.

Weaver, R.D. (1996) Prosocial behavior: private contributions to agriculture's impact on the environment. *Land Economics* 72(2):231–247.

Weersink, A., Livernois, J., Shogren, J.F. and Shortle, J.S. (1998) Economic instruments and environmental policy in agriculture. *Canadian Public Policy* XXIV: 309–327.

Weinberg, A.C. (1990) Reducing agricultural pesticide use in Sweden. *Journal of Soil and Water Conservation* 45:610–613.

Weinberg, M. and Kling, C.L. (1996) Uncoordinated agricultural policy making: an application to irrigated agriculture in the West. *American Journal of Agricultural Economics* 78:65–78.

Weinberg, M., Kling, C.L. and Wilen, J.E. (1993a) Water markets and water quality. *American Journal of Agricultural Economics* 75:278–291.

Weinberg, M., Kling, C.L. and Wilen, J.E. (1993b) Analysis of policy options for the control of agricultural pollution in California's San Joaquin Basin. In: Russell, C.S. and Shogren, J.F. (eds) *Theory, Modeling and Experience in the Management of Nonpoint-Source Pollution.* Kluwer Academic Publishers, Dordrecht, The Netherlands.

Weitzman, M. (1974) Prices vs. quantities. *Review of Economic Studies* 41:477–491.

Wetzstein, M.E. and Centner, T.J. (1992) Regulating agricultural contamination of groundwater through strict liability and negligence legislation. *Journal of Environmental Economics and Management* 22:1–11.

Whitehead, J.C. and Blomquist, G.C. (1991) Measuring contingent values for wetlands: effects of information about related environment goods. *Water Resources Research* 27:2523–2531.

Wicker, W. (1979) Enforcement of Section 208 of the Federal Water Pollution Control Act amendments of 1972 to control nonpoint source pollution. *Land and Water Law Review* 14(2):419–446.

Williams, J.R., Jones, C.A. and Dyke, P.T. (1984) A modeling approach to determining the relationship between erosion and soil productivity. *Transactions of the American Society of Agricultural Engineers* 1: 129–144.

Williams, J.R., Nicks, A.D. and Arnold, J.G. (1985) Simulator for water resources in rural basins. *Journal of Hydrology Engineers* (American Society of Civil Engineers) 111:970–986.

Willig, R. (1976) Consumer's surplus without apology, *American Economics Review* 66:589–597.

Wischmeier, W.H. and Smith, D.D. (1978) *Predicting Rainfall Erosion Losses – a Guide to Conservation Planning.* Agriculture Handbook 537. US Department of Agriculture, Science and Education Administration, Washington, DC.

Wolf, S.A. and Nowak, P.J. (1996) A regulatory approach to atrazine management: evaluation of Wisconsin's groundwater protection strategy. *Journal of Soil and Water Conservation* 51(1):94–100.

World Bank (1992) *World Development Report 1992: Development and the Environment.* Oxford University Press, New York.

World Health Organization (1990) *Public Health Impacts of Pesticides Used in Agriculture.* World Health Organization, Geneva.

World Resources Institute (1996) *World Resources 1996–97.* Oxford University Press, New York.

Wu, J. and Babcock, B. (1995) Optimal design of a voluntary green payment program under asymmetric information. *Journal of Agricultural and Resource Economics* 20:316–327.

Wu, J. and Babcock, B. (1996) Contract design for the purchase of environmental goods from agriculture. *American Journal of Agricultural Economics* 78 (Nov):935–945.

Wu, J. and Babcock, B. (1999) The relative efficiency of voluntary vs mandatory environmental regulations. *Journal of Environmental Economics and Management* 38:158–175.

Wu, J. and Segerson, K. (1995) The impact of policies and land characteristics on potential groundwater pollution in Wisconsin. *American Journal of Agricultural Economics* 77(4):1033–1047.

Xepapadeas, A. (1991) Environmental policy under imperfect information: incentives and moral hazard. *Journal of Environmental Economics Management* 20:113–126.

Xepapadeas, A. (1992) Environmental policy design and dynamic nonpoint source pollution. *Journal of Environmental Economics Management* 23:22–39.

Xepapadeas, A. (1994) Controlling environmental externalities: observability and optimal policy rules. In: Dosi, C. and Tomasi, T. (eds), *Nonpoint Source Pollution Regulation: Issues and Policy Analysis.* Kluwer Academic Publishers, Dordrecht, The Netherlands.

Xu, X. (1999) Do stringent environmental regulations reduce the international competitiveness of environmentally sensitive goods? A global perspective. *World Development* 27:1215–1226.

Yiridoe, E.K. and Weersink, A. (1998) Marginal abatement costs of reducing ground water-N pollution with intensive and extensive farm management choices. *Agricultural and Resource Economics Review* 27:169–185.

Young, D.L., Walker, D.J. and Kanjo, P.L. (1991) Cost effectiveness and equity aspects of soil conservation programs in a highly erodible region. *American Journal of Agricultural Economics* 73(4):1053–1062.

Young, R.A., Onstad, C.A., Bosch, D.D. and Anderson, W.P. (1987) *AGNPS, Agricultural Nonpoint Source Pollution Model: a Watershed Analysis Tool.* Research Report No. 35. US Department of Agriculture, Agricultural Research Service, Washington, DC.

Zilberman, D., Khanna, M. and Lipper, L. (1997) Economics of new technologies for sustainable agriculture. *Australian Journal of Agricultural and Resource Economics* 41:63–80.

Index

Figures in **bold** indicate major references.
Figures in *italic* refer to diagrams, photographs and tables.

acreage restrictions 14
ad valorem tax 159
administration costs 121
aesthetic damage 5
after-tax profits 28, 29
agricultural agencies 17
agricultural extension programmes 69
agricultural externalities 14
    control of 162
    domestic 164
    reducing 67–84
agricultural load reductions 12, 14
Agricultural Nonpoint Source model
    (AGNPS) 88
agricultural policy distortions 116
agricultural pollutants 3–9, 39–43, 54
    chemicals 130
    environmental impacts 9–12, *11*
    protection of surface water 143
    reducing 12–17
agricultural production, changes in
    112–115
Agriculture, Conservation, and Trade Act
    (USA) 133
Agri-Environmental Regulation (EU)
    151, 152, 153, 161
Albemarle, USA 5
algae growth 4, 144
altruistic producers 72–73

ambient pollution concentrations 22,
    23, 26–27, 30–35, 51–52
ambient pollution taxes 49–52
amenity value 106
animal manures, application 152, 160
animal production *see* livestock
animal waste 5, 9, 141, 145–146
aquatic ecosystems 4, 5, 7, 124
Aral Sea, Central Asia 8, 9
asymmetric information 46–48, 51
Atchafalaya River, USA 6
atmospheric deposition (nitrogen) 4, 6
atrazine 139–140, 154
Australia 24, 172

bacteria 9
Baltic Sea 157
banded tax system 156
basin models *see* watershed models
behavioural responses 86–87, *88*, 89,
    99
benefit/cost ratio 83
benefit–cost linkages *86*
best management practices (BMPs) 129,
    136–137, 145, 146
biomagnification 7
biotechnology 75
bird deaths 1

blue-baby syndrome 6
BOD (biological oxygen demand) 154,
    156–157
Bothnian Bay 157
buffer strips 77, 157

California, USA 44, 137–139
Canada 24, 54, 69, 172
cancer 6, 7
carcinogens 7
Central Platte, Nebraska, USA 140
Central Valley Project Improvement Act
    (USA) 9
centralized policies 14, 15, 135
chemical indicators 89
chemicals 21, 130
Chemicals, Runoff and Erosion from
    Agricultural Management
    Systems (CREAMS) 88
Chesapeake Bay, USA 5, 12, 15, 82, 84,
    87, 88, 98, 135, 142
chicken production 5
Chile 54
choice experiments 110–111
Clean Air Act 14, 123–126, *125*,
Clean Water Act (USA) 14–15, 38, 123,
    127–129, 134–135, 145
Clean Water Action Plan 68, 147, 148
coastal waters 5, 157
Coastal Zone Act Reauthorization
    Amendments (CZARA) 129
Colorado River, USA 127
command-and-control instruments 13,
    15, 157
commodity market distortions 60–61
commodity price boom 82
Common Agricultural Policy (CAP), EU
    152–153, 161
compensating surplus 94
compensating variation 94
compliance measures 28, 82–84, 117,
    159
Comprehensive Environmental Response,
    Compensation, and Liability Act
    (CERCLA) (USA) 131
Comprehensive Nutrient Management
    Plans (CNMPs) (USA) 147
comprehensive policies 74, 78

computable general equilibrium (CGE)
    172, 175, 180
concentrated animal feeding 5,
    145–146
conservation compliance 82–84
Conservation Compliance programme
    (USA) 82, 132
Conservation Reserve Program (CRP)
    (USA) 38, 79, 83, 111, 133
Conservation Security Act (USA) 79
conservation tillage 69
consumers 93–94, 97, 99, 109,
    115–116
consumption 168–169, *170*, 176
    environmental services 99–111
contingent valuation 108–110, 112
contracts 22, *23*, 24
coordination of policies 28, 120, 161
Corn Belt (USA) 45, 82, 83
cost estimation
    lessons from 119–121
    pollution control 120–121
    producer–consumer costs 115–116
    production changes 112–115,
        114, 117–119
    social costs 115–116
        measurement of 116–117
cost/benefit estimation
    basic concepts and procedures 85
    benefit–cost linkages *86*
    environmental policy impacts
        86–92, *86, 88, 90, 91, 92*
    valuation 92–94
Costa Rica 175, 180
cost-effectiveness 12, 15, 28, 162
    green payments 79–80, 81
    point/non-point trading 55, 59
    policies 30–38, 44–45, 120, 149,
        163–164
    taxes 49–52, 157
costs
    administration 121
    agricultural support programmes
        67
    changes in economic activity 16
    compliance 117–118
    crop production 83, 138
    dead weight 16, 61, 116, 120
    education 68

environmental damage 26–27,
112, 156
environmentally friendly practices
13, 77, 135–136
health 127
inputs 28
marginal 96
of obtaining information 16
pollution controls 27, 38, 60, 117,
120–121
social 28, 115–117, 177–178
transactions 59
cost-savings 96
Council of Ministers (EU) 151
Countryside Act (UK) 153
crop rotations 74
cross-compliance 161
Cryptosporidium 9, 127

dairy farming 9, 141, 154
dam building 2, 9
damage cost function 32
DCBP (nematocide) 7
DDT (pesticide) 7
dead zones 6
decentralization 15, 16
defensive expenditures 99–101
Denmark 24, 69, 157, 158
Pesticides Action Program 159
derogations 152
developed countries 164, 171, 172,
174, 179
developing countries 174, 179
direct compliance costs 117–118
directives (EU) 151–152
discharge-based economic incentives 25
distortionary taxes 159
downstream jurisdictions 15
drinking water 6, 100–101, 111
contamination 11
nitrate levels 155, 126–127
pathogen 9
pesticides 7
Drinking Water Contaminant Candidate
List (USA) 130
Drinking Water Directives (EU) 151, 155
dynamic input-based instruments
36–37

East Anglia (UK) 152
eco-imperialism 164
economic activity
mix of 166, 167–175
scale effects 165, 166, 167–175
economic benefits 93–94, 111
economic costs 126
variables 85
economic criteria 19, 162
economic incentives 13, 53–54
economic modelling 118–119
economic responses, new technologies
75–77
economic thresholds 74
economically efficient level (water
quality) 70–71
economically preferred base 21
economic theory (non-point pollution
control) 25–26
ecosystem maintenance 9
eco-taxes 158–159
EDB (nematocide) 7
education 68–69
comprehensive policy 74
producers and environment 72–74,
72
producers and profitability 69–71,
70
egg production 5
emission proxies 21, 23, 26–29
regulation 37–44, 39–43
emissions estimate 26
emissions flow 21
emissions taxes 77
emissions trading 57–60
emissions, stochastic 29–35
emissions-for-estimated loadings (E-EL)
trading 57, 59
emissions-for-inputs (E-I) trading 57,
58–59
enforcement 37, 146
entry and exit 35–36
envelope theorem 95–96, 115
environmental agencies 17
environmental benefits index (EBI)
133
environmental benefits, Nitrate Sensitive
Areas (NSA) 155
environmental capital 169–170

environmental damage
    costs 26–27, 32, 112
    impacts of
        location 15–16
        on people *11*, 12, 16
    indicators 27
    trade-offs 81
environmental externalities 164
Environmental Frameworks (EU) 151
environmental impact coefficients 157,
    158
Environmental Law Institute 17
environmental outcome targeting 162
environmental policy instruments
    13–14, 19–20, *23*, 120
    centralized 135
    choice of base 29, 33, **45–48**
    cost/benefit estimation *see*
        cost/benefit estimation
    design of 60–61
    impacts of 13–16, 86–92, *86*, *88*,
        90–91, 92
    incentives and regulations 26–29
    international coordination
        163–164
    R&D 77–78
    recommendations 61–62
    tool kit 22–26
    what stimulus? 22
    what to target? 21–22
    whom to target? 20–21
Environmental Protection Act (UK) 155
environmental protection, budgets for
    153
environmental quality 27, 93, *98*, 106,
    107
environmental services 99
    choice experiments 110–111
    contingent valuation 108–110
    defensive expenditures 99–101
    generalized travel cost analysis
        105–106
    hedonic pricing 106–108
    hedonic travel cost analysis 104
    site choice models 103–104
    travel cost 101–103
environmental side agreement 179
environmental state variable 22
environmental taxation 160

environmentally friendly technologies
    13, 76–77, 180
environment-constant trade-off
    170–171
Environmental Quality Standards (UK)
    154
Environmentally Sensitive Areas scheme
    (UK) 157, 161
equity objectives 80
Erosion-Productivity Impact Calculator
    model (EPIC) 88, 89
Estonia 158
estuaries 4, 9, 124, 142
European Commission 151
European Union (EU) 151–153,
    160–162
    land rents 181
    Scandinavia *see* Scandinavia
    The Netherlands *see* Netherlands,
        The
    trade liberalization 168, 175, 180
    United Kingdom *see* United Kingdom
eutrophication 4–5, *11*, 157, 158, 159
evaporation (pesticides) 7
Everglades, Florida, USA 141–142
expected water quality 69–70, *70*, 73
expenditures 94
export subsidies 152
externalities *see* agricultural externalities
externality effects 167

fairness 20
farm commodity programme (USA) 82,
    83
farm production practices 26–29
    regulation 37–44, *39–43*
farmers
    education 68–69
    health risks 10
    perceptions of water quality 73–74
    pollution prevention 19, 113
        payments 117, 152, 153, 154,
            155, 156
    property rights 153, 154
Farmers Home Administration (FmHA)
    benefits (USA) 83
farm-level mineral accounting system
    159

farms
   costs of pollution control 12, 13
   environmental performance
      indicators 21
   estimated runoff 28–29, 33–35
   income 79, 81, 83, 156
   land value 10
   manure 154, 155, 160
   metering pollutant flows 20
   R&D 78
farm-specific environmental performance
      indicators 21
farm-specific green payments 81
Federal Insecticide, Fungicide and
      Rodenticide Act (FIFRA) (USA)
      131
Federal Water Pollution Control Act
      (USA) *see* Clean Water Act (USA)
federalism, economic theory 14, 135
fertilizers 3, 20, 156, 157
   application limits 160
   reduction 120–121
   taxes 45, 113–114, 115–116,
      117–118
      Scandinavia 157, *158*
   tradeable permits 116
field-scale models 88–92
Finland 157, 158, 161, 162
first-best policies 37, 45–46, 50
fish 1, 4, 8
fisheries 5, 8, 89, 127
flexibility (policies) 163, 164
Florida, USA 24, 141–142, 148
Food Security Act (USA) 83
fossil fuels 6
France 162
freshwater
   eutrophication 4
   quality 89, 92, *92*
   recreation 111–112, 127

gastrointestinal illness 9, 127
genetically modified organisms (GMOs)
      75
Germany 54, 158
Giardia spp. 9, 127
governments
   agencies 17, 78

guidance 128, 129
intervention 60, 152–153, 160
national 15, 16
   Scandinavia *see* Scandinavia
   The Netherlands *see*
      Netherlands, The
   United Kingdom *see* United
      Kingdom
   USA *see* USA
revenue 179
grains 175
grazing pressures 161
Great Lakes Region (USA) 82, 83
Great Plains (USA) 38, 82, 83
green payments 79–82
Grossman, Gene 165
groundwater 112, 124
   pollution 1–2, 7, 126, 129,
      130–131
   atrazine 139–140
   impacts of 15
   nitrogen 140–141
   pesticides 137–139
Groundwater Loading Effects of
      Agricultural Management
      Systems model (GLEAMS) 88
gulf hypoxia 87
Gulf of Mexico (USA) 3, 4, 6, 15, 82,
      84, 120, 135

habitats 1, 8, 133
hedonic pricing 106–108
herbicides 139, 156
Hicksian compensating measure 93–94,
      97, 99
highly erodible land (HEL) 83
Hill Livestock Compensatory Allowances
      (UK) 161
horticultural industry 159
human capital 173–174
human health risks 6, 7, 9, 10, 16, 127
human waste 9
Hurricane Floyd 5
Hydrologic Simulation Program (HSPF)
      88

Illinois, USA 146

import tariffs 179
incentives 19, 22, 26–29, 48–49,
        77–78
    ambient-based 52–54
    application 37–44, *39–43*
    economic 24, 157, 158
    uniform versus differentiated
        44–45
income 92–93
    changes in 95
    farm 79, 81, 83, 156
    government 179
    marginal utility 178–179
indirect approaches 68
Indonesia 175
industrial pollution 128, 171–172
inorganic fertilizers 156
input mix effects 167, 173–175, *175*
input permits 59
input subsidies 13
input suppliers (R&D) 78
input-based incentives 24–25
input-based instruments 29, 45–46
    mixed 48
inputs 26–29, 76–77
    fate of 29–35
    regulation 37–44, *39–43*
    taxes 113
    trading 55, 57
insecticides 159
integrated pest management (IPM) 69,
        71, 74, 75, 156
international involvement 15–16,
        163–164
international trade 81–82
intervention, government 160
Iowa, USA 24
irrigation 1, 8–9, 10, 44, 45, 69
isoproturon 154

Japan 168, 172
jurisdictions 15–16

Kattegat Strait 157
Kentucky, USA 144–145
Kesterson Reservoir, California 1, 8
Krueger, Alan 165

Lake Okeechobee, Florida 24, 141, 142
Lake States, USA 45
lakes 15, 124
land rents 181
land retirement programmes 133
land use, change of 24
landscape 152, 153, 156, 162
Latvia 158
leaching 4, 44
least-cost solution 28
legislation 164
lettuce production 44
liability rules 22, 25, 26, 52–54
licensing, pesticides (UK) 156
Lithuania 158
litigation 54
livestock
    dairy farming 9, 141, 154
    headage payments 153, 161
    production 3, 5, 14, 22, 131–132
local authorities 14
local solutions 15
low level exposure (pesticides) 7
lump-sum payments 36
lump-sum transfers 80

Maastricht Treaty 161
management action targeting 162
Management Agreement model 153,
        154
management plans 129
manure spreading 154, 155, 160
manure trading system 160
marginal changes 111
marginal costs 61, 136, 156
    social 117, 177–178
marginal cost-savings 96
marginal impacts, nitrates *158*
marginal implicit price 106–107
marginal rate of substitution (MRS) 72,
        73
marginal trade-offs 80–81, 171–172,
        171, 172
marginal value 103, 104
market-based approaches 13
markets *23*, 60–61
Maryland, USA 142
mass balance approach 159

mathematical models, cost estimation 118–119
metered discharges 20, 21
methaemoglobinaemia *see* blue-baby syndrome
Mexico 172, 182
Midlands, UK 152
mineral damage 8–9
Ministry of Agriculture (MAFF), UK 153, 154
Miranowski, John 165
Mississippi River, USA 3, 6, 120
mix effects 166
multiple objectives 67–84
Murray-Darling, river, Australia 8, 9

National Estuarine Eutrophication Survey (USA) 4
National Oceanic and Atmospheric Administration (NOAA) (USA) 129
National Oceanographic and Atmospheric Administration (NOAA) (USA) 110
National Oceanographic and Atmospheric Administration (NOAA) (USA) 4
National Parks and Access to the Countryside Act (UK) 153
national pesticide regulation 15
National Pollutant Discharge Elimination System (NPDES) (USA) 128
National Rivers Authority (UK) 154
Nature Conservancy Council (UK) 153
navigation channels 126
Nebraska, USA 140–141
negligence 53–54
nematocides 7
Netherlands, The 3, 5, 24, 46, 69, 159–160, 162
new technologies 75–77
New Zealand 172
Nitrate Sensitive Areas (NSA) 155
Nitrate Vulnerable Zones (NVZs) 152, 159
nitrates **151–152**
    drinking water standards 6
    loss 44, 45, 49, 159

marginal impacts *158*
pollution 48, 140, 157
taxes 158
Nitrates Directive (EU) 151–152, 155, 159, 162
nitrogen 44
    fertilizer 20, 157
    protection of groundwater 140–141
    tax 61
    water pollution 3–6
Nitrogen Leaching and Economic Analysis Package model (NLEAP) 88
nitrogen-fixing cereal varieties 75
non-distortionary policies 80
non-economic criteria 19
non-point emissions 26
non-point production function 26
non-point production function 29–30
non-point source pollution 2–3, 4, 5–6, 15–16, 18, 148–149
    assigning responsibility 20
    conservation compliance 82
    control 22–26
    environmental instruments 28–29
    policy design 20, 134–136, 163–164
        economics of 21, 26–29
    Scandinavia 157
    USA *see* USA
normative estimating methods 98, 118–119
North American Free Trade Agreement (NAFTA) 163, 164, 179, 182
North Carolina, USA 5, 143–144
Norway 157, 161, 162
no-till farming 10, 49, 76
nutrients
    levels 156
    management 69
    pollution 3–6, 126, 142
    protection from 142
    reduction 88

off-farm environmental impacts 10–12, *11*
on-farm environmental impacts 9–10

Organic Aid scheme (UK) 156
organic fertilizers 156–157
organic pesticides 6–7
Organization for Economic Cooperation
    and Development (OECD) 4–5,
    67, 165, 166, 167, 177
output mix effects 167–173
oxygen deficiencies 6

Pareto efficiency 70, 162
partial budgeting 117
participation rates 153, 154
pathogen damage 9
'pay the polluter' principle 79
payment rates 161
payment schemes
    area-based 153
    per hectare 155, 156
    voluntary 161
per capita income 178–179
performance indicators 21
permits 22, 47, 57
    fertilizer 116
    input 58, 59
    point source discharge 144
    pollution 54, 77
Pesticide Root Zone model (PRZM) 88
pesticide tax
    Scandinavia 158–159
    UK 156
pesticides 6–7
    accumulation 7
    groundwater 126
        protection 137–139
    national regulation 15
    registration 22, 24, 152
    use (UK) 154
    water pollution 3, 127, 155
Pesticides Action Program (Denmark)
    159
phosphate pollution, Scandinavia 157
phosphorus
    sedimentation 8
    surface water protection 141–142
    water pollution 3–6
physical capital 173–174
point source controls 15
point source emissions 26

point source pollution 2, 4, 148
Point Source Program (USA) 128
point/non-point trading 54–57, 56
Poland 158
policies, efficiency of 14, 16
policy coordination 28, 120, 161
policy cost estimates 119–121
policy effects 167
political support (of regulation) 20
'polluter pays' principle 79, 151, 162,
    164
pollution abatement 76, 77
pollution control instruments *23*
pollution control policies *see*
    environmental policy
    instruments
pollution flows 26, 87
pollution haven hypothesis 179
pollution load reductions 12
pollution prevention 76
pollution sources 2–6, 12
pollution trading 25, 54–57, 56
positive estimating methods 98,
    118–119
prevention action limit (PAL) 139
price and income policies 13
price elasticity 118, 156
price supports 13, 175
price vector 93
prices
    agricultural goods 61, 168, 174
    changes 95–98
    guaranteed 152
    inputs 24, 87
    output 87
    producer 13, 45
probability density function (pdf)
    30–32, *31*
probability sampling 110
producers
    altruistic 72–73, *72*
    costs 115–116
    improving environmental
        performance 22
    profit-maximizing 69–71, *70, 76*
    regulation 20–21
    small 68
product charges 156
product effects 165, 166

production 167–169, *169*, 176
    agricultural 112–115
    optimal sites 36
production decisions 30–31, 70–71, *70*
production frontier 71
production possibilities frontier (PPF)
        167–168, *169*, 170, *170*, 171,
        171, *172*, 176, *177*
production quotas 14
productivity improvements 95–98
profits 28–29
    maximizing 69–71, *70*, 95–96
    reduction 116
property rights 153
property value 106–108
protozoan parasites 9
proxies 46
    emissions 21, *23*, 26–29, 37–44,
        39–43
    farm level environmental
        performance 113–115
public agencies 68
public persuasion 22

quantity controls 47, 49
Quebec, Canada 5

recreation 5, 27, 89, 103, 111–112,
        153
recycling rates 159
regulations 24–25, 26–29
    application 37–44, *39–43*
        preferred use 48–49
    direct 22
    EU 151,152
    pesticides 13
    USA 128
regulatory effects 165, 166
regulatory standards 21–22
research and development (R&D) 74–75
    comprehensive policy 78
    incentives for 77–78
reservoirs 1, 8, 15, 126
Resource Conservation and Recovery Act
        (RCRA) (USA) 130–131
risk effects 46, 51
rivers 4, 5, 124

Roy's identity 168
run-in 3–4
runoff 3, 10, 30, 45
    dissolved pesticides 7
    estimated 28–29, 33–35, 45–46

Safe Drinking Water Act (SDWA) (USA)
        14, 127, 129
safety-first approach 30–32, *31*
salinity levels 8, 10, 127
salinization 9–10
scale effects 165, 166, 167–175
Scandinavia
    pesticides 158–159
    pollution from nitrates and
        phosphates 157–158, *158*
seafood 5
Seattle, USA 163
second-best policies 37–38, 45, 60–61
sediment 3, 4
sedimentation 7–8, *11*, 126
shellfish 7, 9
silage effluent 154, 155
Silage, Slurry and Fuel Oil Regulations
        (EU) 155
Simulator for Water Resources in Rural
        Basins model (SWRRB)
site choice 36, 103–104
Sites of Special Scientific Interest (SSSI)
        153
Small Watershed Program 133
social benefits 97
social costs 28, 115–117, 177–178
social objective function 80, 81
Soil and Water Assessment Tool (SWAT)
        87
soil erosion 4, 7, 10, 112, 133
    no-till farming 10, 49, 76
    predicting 21, 89
    reducing 83, 111
soils 44
Sole Source Aquifer Protection Program
        (USA) 130
Source Water Assessment Program
        (SWAP) (USA) 130
SPARROW (Spatially-Referenced
        Regression on Watershed
        Attributes) 87

spillovers 15, 16, 176–177, 180
Sri Lanka 175
standards *23*, 24, 61
staple crops 172
State programmes (USA) 136–145, *138*
stewardship 72
stochastic emissions 29–35
Stolper–Samuelson considerations 174
stormwater 142
streams 4, 15
stressors 89, *90*
structural effects 165, 166
Structures Regulation (EU) 153, 160
sub-national authorities 15, 16
subsidiarity principle (EU) 151
subsidies 22, *23*, 24–25, 115–116
    export 152, 179
    input 13, 47, 80, 113
    lump-sum 36
    national 16
    optimal rates 80–81
    price 116–117
    UK 154
supply controls 14
support payments (agriculture) 79
surface water
    monitoring 124
    pollution 1, 2, 124
        pesticides 7, 127
    protection
        agricultural pollution 143
        phosphorus 141–142
    quality *125*, 129
Swampbuster programme (USA) 82
Sweden 24, 69, 157, 158, 162

targeting 119–120, 161–162
Tar-Pamlico Basin, North Carolina 5, 144
taxes 22, *23*, 24, 142, 179
    ad valorem 159
    ambient pollution 49–52
    distortions 16, 61, 159
    eco 158–159
    environmental *114*, 160
    farm manure surpluses 160
    farm specific 37
    fertilizer 45, 113–114, 115, 157,
        158

income 159
inputs 44, 47, 78, 113, *114*
lump-sum 36
nitrate 158
pesticide
    Scandinavia 158–159
    UK 156
rates of 29, 113–115
    efficient 26, 28–29, 33, *158*
    optimal 45
    runoff estimates 28–29, 33–35
technical assistance 22
technology effects 165, 166, 167
technology, new *see* new technologies
The Sound, Sweden 157
time horizons 165
total maximum daily loads (TMDL) 55,
    128, 143–144
trade liberalization 163, 164, 168, 172,
    174–175, 179
trade, effects on environment 163–164
    decomposition frameworks
        alternative 166–167
        existing 164–166
        externality effects 176–177,
            177
        input mix effects 173–175,
            175
        multi-country considerations
            182
        policy effects 177–179
        scale and output mix effects
            167–173, *169, 170, 171,*
            172
        technology effects 180–181
tradeable permits 22, 47, 54–55, 77,
    116
trade-offs 80–81, 111, 170–171
trading ratio 57–58, 59, 60
transactions costs 59
transport 29–35
travel costs 101–106, 111
turbidity 7–8, *11*

Underground Injection Control Program
    (USA) 130
Unified National Strategy for Animal
    Feeding Operations (USA) 147

United Kingdom 153–156
 nitrates 152
 pesticide tax 156
 policy initiatives 156–157
 upland regions 161
 water pollution legislation 154,
  155
 water quality 154, 156–157
Universal Soil Loss Equation (USLE) 21,
  88, 89
Upper Mississippi Basin, USA 6
upstream jurisdictions 15
Uruguay Round (trade negotiations)
  163, 164, 172
Uruguay Round Agreement on
  Agriculture (URAA) 81–82
US Department of Agriculture (USDA)
  83, 89, 126
US Department of Interior (USDI) 126
US Environmental Protection Agency
  (USEPA) 1, 7, 14, 15, 89
 National Survey of Pesticides in
  Drinking Water Wells 126
 Water Quality Inventory 124
US Geological Survey 87
US National Pollution Discharge
  Elimination System (NPDES)
  25
USA (United States of America) 1, 24,
  54, 69, 168, 182
 conservation compliance 83–84
 non-point source pollution control
  24
  Clean Water Action Plan 147
  enforcement 146
  evolution of policy 134–136
  state programmes 136–146,
   138
 pollution flow models 87–89, *88*
 trade liberalization 172, 175
 water quality 123–126, *125*, 24
  costs of impairments
   126–127
  protection of 127–133
USDA NRCS Conservation Technical
  Assistance (CTA) program 22
USDA NRCS Environmental Quality
  Incentives Program (EQIP) 22,
  131

USDA NRCS EQIP and Conservation
  Farm Option ((CFO) programmes
  24
USDA Water Quality Projects 73, 133
utility function 177, 178
utility maximization *72*, 93

valuation 92–94
Vermont, USA 143
voluntary compliance approaches 13
voluntary participation 153, 154, 160
voluntary policy instruments 67–68

Waquolt Bay, USA 89
wasteload allocations (WLA) 128,
  143–144
wastewater 123
water pollution
 control programmes 24
 legislation 154, **155**
Water Quality Act (WQA) (USA) 135
water quality improvements
 benefits of 89, 92, *92*
  changes in direct consumption
   99–111
  productivity improvement
   95–98, *96*, *97*
 examples of benefits 111–112
Water Quality Incentives Program
  (WQIP) (USA) 38, 79, 132
water quality, 1–3, 47, 152
 CRP benefits 83–84
 downstream 16
 economic activity 16
 farmers' perceptions 73–74
 farming's impact (UK) 154,
  156–157
 impairment categories 124
 indicators 89
 mineral damage 8–9
 nutrient pollution 3–6
 off-farm environmental impacts
  10–12, *11*
 on-farm environmental impacts
  9–10
 pathogen damage 9
 pesticides 6–7

water quality *continued*
    profitability 69–71
    R&D 78
    reducing agricultural pollution
        12–17
    sedimentation and turbidity 7–8
    USA 123–126, *125*
        impairments 26, 126–127
        protection of 127–133
Water Resources Act (UK) 155
water treatment industry 126–127
waterborne diseases 127
waterlogging 10
watershed models 87–88, *88*, 89,
    90–91, 118
watershed-based management 55, 119
watersheds 5, 6
    export of agricultural chemicals *88*
    pollution 26
    pollution load reductions 12
    risk assessment *91*

watershed-specific information 16
weather 29, 80
welfare economics 80, 97, *114*, 115,
    121
Wellhead Protection Program (WHP)
    (USA) 129–130
wetlands 82, 121, 141–142
Wetlands Reserve Program (USA) 133
wildlife 152, 153, 156, 162
Wildlife and Countryside Act (UK)
    153
willingness to accept payment (WTA)
    94, 108
willingness to pay (WTP) 93, 108–109
win–win opportunities 69, 73, 83,
    120
Wisconsin, USA 139–140
World Health Organisation (WHO) 6,
    155
World Trade Organisation (WTO) 82,
    163, 179

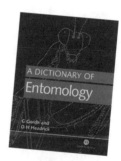
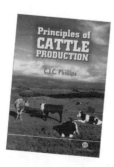